Compulsive Body Spaces

Compulsive Body Spaces presents a spatial understanding of compulsion. Providing a compelling account of the lives of 15 people with Tourette syndrome, it demystifies the seemingly irrational, purposeless and meaningless character of this behaviour.

It demonstrates how attending to the spatial circumstances under which compulsive acts, like touching, ordering, and aligning objects take place, can produce valuable novel insights that complement neuroscientific, psychiatric or psychological knowledge. By paying attention to the sensory, material, and social environment of the body during its performance of compulsive acts, the book establishes the ways in which configurations of bodies, objects, and spaces disrupt people's lives or allow them to thrive. This collaborative, qualitative study that is based on in-depth interviews, observations, and mobile eye-tracking places the book at the forefront of a new wave of patient emancipation in medical research, and gives rise to a renewed consideration of what empathetic, context-sensitive care may look like in the 21st century. In turn, its insights give rise to a ground breaking spatial conceptualisation of wellbeing. Considering the compulsive capacities of a broader humanity, *Compulsive Body Spaces* highlights the compulsive dimension in bodily spatiality, which underpins the very core theories of human life as embodied and performed.

This book will be of interest to students and scholars in science and technology studies, human geography, sociology, health and social care, medical humanities, continental philosophy and disability studies.

Diana Beljaars is a research fellow at the Swansea University Geography Department. Interested in culture, disability, and health, she combines human geography, medical humanities, continental philosophy, and Tourette syndrome-related neuropsychiatry. She has published in *Transactions of the Institute of British Geographers* and co-edited *Civic Spaces and Desire* (Routledge).

Routledge Research in Culture, Space and Identity
Series Editor: Peter Merriman

The *Routledge Research in Culture, Space and Identity Series* offers a forum for original and innovative research within cultural geography and connected fields. Titles within the series are empirically and theoretically informed and explore a range of dynamic and captivating topics. This series provides a forum for cutting edge research and new theoretical perspectives that reflect the wealth of research currently being undertaken. This series is aimed at upper-level undergraduates, research students and academics, appealing to geographers as well as the broader social sciences, arts and humanities.

Geopoetics in Practice
Edited by Eric Magrane, Linda Russo, Sarah de Leeuw and Craig Santos Perez

Space, Taste and Affect
Atmospheres That Shape the Way We Eat
Edited by Emily Falconer

Geography, Art, Research
Artistic Research in the GeoHumanities
Harriet Hawkins

Creative Engagements with Ecologies of Place
Geopoetics, Deep Mapping and Slow Residencies
Mary Modeen and Iain Biggs

Comics as a Research Practice
Drawing Narrative Geographies Beyond the Frame
Giada Peterle

Compulsive Body Spaces
Diana Beljaars

For more information about this series, please visit: www.routledge.com/ Routledge-Research-in-Culture-Space-and-Identity/book-series/CSI

Compulsive Body Spaces

Diana Beljaars

Routledge
Taylor & Francis Group

LONDON AND NEW YORK

First published 2022
by Routledge
2 Park Square, Milton Park, Abingdon, Oxon OX14 4RN

and by Routledge
605 Third Avenue, New York, NY 10158

Routledge is an imprint of the Taylor & Francis Group, an informa
business

British Library Cataloguing-in-Publication Data
A catalogue record for this book is available from the British Library

Library of Congress Cataloging-in-Publication Data
A catalog record has been requested for this book

ISBN: 978-0-367-62608-2 (hbk)
ISBN: 978-0-367-62609-9 (pbk)
ISBN: 978-1-003-10992-1 (ebk)

DOI: 10.4324/9781003109921

Typeset in Times New Roman
by KnowledgeWorks Global Ltd.

Dedicated to Laura Beljaars, Jacques and Wies Beljaars, and Mien Van den Berg-Coppens

"...for in each moment blessed with colourful teachings..."

John Clayton, summer 2016

Contents

Preface

Seldomly have I lived through such a life-affirming moment as I did during that afternoon in the corner room of the 21st floor of the Van Unnik building on the Utrecht University campus. This *truly seismic* shift was encapsulated in a moment that could not be more mundane, and I am pretty sure that none of the other people in the room even noticed it. For their presentation, two of my fellow research master students showed the 2007 YouTube video "In my Language"[1] by non-verbal autistic activist Mel Baggs ('SilentMiaow'). With utter astonishment, I recognised my sister's compulsions in their movements. Also their critique on society's treatment and understanding of their movements, preferences, and life absolutely blew my mind and resolutely recomposed it into a different understanding of the world differently. It came with some sort of sense of vision about my little place in that world which has accompanied me ever since. With the unrelenting ambition that comes with being an early Millennial, I was going to understand compulsion in Tourette syndrome through human geographical theories. Whilst I'm not sure if I have, in fact, improved my understanding of the phenomenon, I am sure that this book brings forth a nuance that starts to do justice to the complexities of the experience and the socio-material circumstances under which they take place.

And that is what this book is about: compulsive *interactions*, not compulsive *people*. Nor is it about Tourette syndrome, or life with Tourette's per se. Equally, I do not claim to have any knowledge about how it is to live with compulsive tendencies as part of a diagnosis. That is mainly because I do not have Tourette's, nor can my proximity to people with Tourette's – my sister most notably – negate this. In fact, I would like to add that such proximity can actually result in a false sense of 'insider' knowledge that, if it goes unchecked, can produce harmful misunderstandings that uphold problematic assertions, reproduce stigma, and negate Tourettic voices.

Whilst this book can be used to identify certain implications of compulsive action for living with Tourette syndrome, it should be read as an outsider analysis of phenomena that can characterise certain aspects of life with compulsivity. As such, this book inhabits the uncomfortably large space left open between patient voices within rare case studies in Tourette's research and in the autobiographies written by people with Tourette's.

To this end, I wrote the book for two main reasons; firstly, because I felt that current-day analyses of *what moves people* are missing the crucial *compulsive* dimension. Therefore, in the book I develop a spatial understanding of compulsivity, and vice versa, a compulsive understanding of living embodied life. And secondly, because people with Tourette syndrome and others with compulsive sensibilities have been, and still are, treated with multiple injustices; some of which seem to go unnoticed. As such, this book pushes for going beyond 'awareness raising' and helps pave the way to a more fundamental understanding of compulsive experience that is needed to develop more socio-cultural empathy, better support structures, and an emancipative shift in how Tourette syndrome and compulsivity are studied.

Diana Beljaars,
Summer 2021

Acknowledgements

Pretty much this entire book has been written in the long isolation that currently continues to be the Covid-19 pandemic, and it is most definitely a product of these circumstances. This feat could not have gained the shape that has solidified in, if it wasn't for countless individuals, groups, and institutions that have granted my work an intellectual home, allowed it time to mature, and were patient with me in my times of stress and frustration.

First and foremost, I have eternal gratitude to my loving parents, Jacques and Wies Beljaars and my amazing sister Laura Beljaars, whose love in its many forms always makes everything possible. I would particularly like to acknowledge the depth with which Laura's critical thoughts, concerns, and beliefs have shaped my thinking about Tourette's and compulsions: to say that her presence from the seedling idea all the way back in 2010 to this manuscript has been incremental is an understatement.

I also very gratefully acknowledge the intellectual homes and resources of the Swansea University Geography Department, the Cardiff University school of Geography and Planning, and the Utrecht University Department of Human Geography and Spatial Planning. I am also very grateful for the ESRC DTP Wales which granted me a Postdoctoral Research Fellowship 2019–2020 (grant code ES/T009268/1), and Swansea University which granted me a five-month UKRI Covid Extension (grant code EP/520597/1).

The unfolding of the manuscript took place in my time at the Swansea Geography Department, under the brilliant mentorship of Marcus Doel. I cannot be more thankful for his intellectual generosity and curiosity, unwavering encouragement in moments of self-doubt, and his comprehensive and strategic guidance. His advice did not only help me navigate writing my first book, it put many a pandemic challenge in perspective, often packaged with a nifty sense of humour. I feel a similarly profound gratefulness to Jo Bervoets for our many long conversations, email exchanges, and collaborative work, which have shaped many of the conceptual and theoretical underpinnings of the work in important ways. I have been very lucky to think and work with him for his nuanced and deep understanding of the multilayered connections between philosophy, the Tourette syndrome-related biomedical and clinical sciences, and human geography. Working with him, Hanne de Jaegher and most recently, Daniel Jones, provided me with the vital sense of community that I sorely missed when I wrote my PhD dissertation.

The book's rooting in my PhD research also allows me to gladly express my enduring immense gratitude for the 15 amazing people who gifted their thoughts, time, and energy for the doctoral research project. I am so thankful for their trust in me to explore their compulsive lives with them and write it up. I am also intensely grateful for the fantastic ongoing guidance and support I have been lucky enough to receive from Jon Anderson, and I am also keen to acknowledge the many forms of support and mentorship of Mara Miele, Julian Brigstocke, and Cara Verdellen. Furthermore, I would like to underscore the crucial encouragement offered by Felicity Callard to develop the dissertation into a monograph. I will never not be endlessly impressed by the productive generosity and criticality of her work across a plethora of disciplines that she so elegantly delivers. It forms a constant source of motivation and patience needed to do the sometimes very challenging work that comes with doing interdisciplinary research, as is her gentle mentorship and leadership through multiple crises, all of which I am deeply thankful for.

My gratitude further extends to my wonderfully supportive colleagues at the Swansea University Geography Department – in particular Osian Elias, Chris Muellerleile, Dave Clarke, Angharad Closs-Stephens, Sergei Shubin, who very kindly arranged the start of the COVINFORM project around the finalisation of the manuscript, Amanda Rogers, Anna Pigott, Katie Preece, Stephen Cornford, Siwan Davies, Rhian Meara, Richard G. Smith, and Bernd Kulessa, as well as to Louisa Knowles for proofreading. Grateful acknowledgements are also due to the Research Hub's Fiona Jones and Sean Richardson, Stichting Gilles de la Tourette including Hans Eijsackers, Sandy Wong, James Ash, Paul Harrison, Paul Simpson, David Bissell, Sarah Atkinson, Angela Woods, Sage Brice, Andy Williams, Martin Dijst, Seonaid Anderson and Lisa Keenan, the AAG Disability Specialty Group board, the "Diverse" AAG Specialty Group Chairs 2020, the ERC NeuroEpigenEthics group, and the many academic others who have been informative and supportive over the last 7 years. And I remain forever grateful for the ongoing mutiplicitous presence of the late John Clayton, whose gentle guidance and playfulness I find pushing my words.

I have an amazing group of absolutely wonderful people around me that I am lucky enough to think with, bask in the presence of, and share good and bad times with. Through the pandemic adversities, these people were particularly amazing, and I am so deeply thankful for each and every one of them: Maartje Kuiper, Lisanne Struckman, Roel Theunissen, Rob Tönissen, Tara Hipwood, Rich Gorman, Jen Owen, and Amy Walker. I'm particularly grateful for Aspasia Karyofili, whose support in the later phases of the project has been absolutely invaluable. I also gladly acknowledge Lia and Cees Nuyten, Charles Drozynski, Kieran O' Mahony, Stephanie Barille, Charlotte Eales, Jack Pickering, Lyndsey Stoodley, Hade and Richard Gale, the Swansea Geography PhD community, the stars of Team South

Wales, as well as Raja, Katya Zamolodchikova, Trixie Mattel, the foxes and other animals in the garden Spring/Summer 2020, Swansea Bay west, and my Leeds volleyball.

Final acknowledgements go to everyone who doubted that the combination of human geography and compulsion and/or Tourette syndrome would form a viable research direction.

Note

1. Complete source: Baggs, M. (silentmiaow). (2007, January 14). *In my language* (Video file). Retrieved from https://www.youtube.com/watch?v=JnylM1hI2jc&t=31s.

Introduction

Hardly a day goes by where we do not confuse ourselves about what we just did and why. This confusion does not pertain to our actions in the grand scheme of things, but to our movements in the lulls and during the mundane tasks in between. Indeed, we have pretty much mastered the former: research can tell us all about how reason, a variety of rationalisations and emotionality governs our lives. The latter, however, not so much. Making sense of what we do is something we seem to do a lot of. Usually we come to a conclusion on why ordering the fridge in this particular way was *helpful*, or why the strictly organised collection of candles, plant pots, and figurines on the windowsill is *meaningful* to us, and why the only *right* place for the remote control is the left arm rest on the sofa. Other times we fail to readily explain what we just noticed ourselves do; stepping over the lines in the pavement, tapping the handle once more after closing the door, creating a balanced pattern of the jars, cups, pots, and pans when we set the table for a meal.

We do not tend to spend much time pondering these moments, despite realising we find ourselves in them more often than we thought if we really pay attention. Nonetheless, when we catch ourselves doing such acts, we cannot quite make sense of how they affect us. They might annoy us for the slight delay they cause, but could also put an extra spring in our step, and add a welcome sense of familiarity to our day. Indeed, such acts could easily embarrass us, but also affirm our beloved quirkiness. In one way or another, these seemingly insignificant practices are affirmative of our character, and reflect the ongoingness of our life. This book dwells on these moments, because fleeting as they are, the sudden vigour with which we feel very strongly about things that do not seems to matter all that much tells us a lot about ourselves, the way we relate to the things around us and the spaces our lives play out in, and ultimately, our place in the world. Whilst pretty much everyone will be familiar with these inexplicable acts, but does not spend much time pondering them, people with Tourette syndrome are intimately knowledgeable of them, as their days can be filled with doing such acts.

DOI: 10.4324/9781003109921-1

It is some 18 years ago. My family has just finished supper. As always, my father, mother, and I get up from our chairs, pick up the glasses, empty pots and pans, and bring them to the kitchen. We all know that my sister, Laura[1], will not pick up the plates with the cutlery on top; not because she does not want to, but because she *just can't*. Albeit with some reluctance, it had made its way into the family psyche: this is simply how it goes. It was normal. In the kitchen she puts the bottles of condiments she had picked up instead on the counter so that they align with each other and the counter edge perfectly. Before turning around to go back, her hand goes towards the bottom left corner of a cupboard door, and she pushes her thumb firmly in its tip. She then notices a drawer that had not shut entirely, upon which she has to close it, despite my mother using it. With every step she takes, she mentally measures where she puts her foot as to not break the lines between the tiles, as well as the imaginary ones extending from the corners of the kitchen and the kitchen table. I had seen it a thousand times, and would see it a thousand times more; Laura had explained a thousand times not to know *why* she did it and simultaneously being unable not to do it. And she would repeat this a million times more in the decades to follow.

She did not choose to do these acts, and they are purposeless. Nonetheless, she *has* to do them in order to leave the kitchen. These household items do not hold a personal meaning for her, nor do the lines that mark the kitchen floor trigger a particular memory, nor a fear, or even an emotion. There is 'simply' *no other way* she can move around in the home and interact with stuff around her. If anything, these practices resemble an economic transaction, or at times more accurately, an extortion. For Laura to continue life as chosen, intended, and as enjoyable, doing these practices as efficiently as possible helped to reduce the unmistakable disruption they incited. Being compelled to disrupt the flow of everyday life to align things, order them in a way that was 'right', and touching things she did not want to touch tormented her. These acts had been accompanied by pulling on clothes to make them feel 'just-right' on her body and sniffing without having a cold. No one understood what was happing to her. *To* her, because she could not explain why she had to do them, nor why she had to in the ways they happened, or why there and then. She only knew that her body needed to do these things because it urged her to; she compares it to having a mosquito bite that itches so badly, you cannot but scratch it. These 'extremely rigid preferences' that guided the ways in which she did things, wanted her room to look, and had her belongings ordered, made her deeply doubt who she was and what she wanted. These acts were voluntary but unavoidable, unexpected, meaningless and irrational, and, as such, they were compulsive. After four gruelling years of seeking the right support, she was diagnosed with of Gilles de la Tourette syndrome (or Tourette's in short) according to the DSM-5 (307.23) (2013) from the American Psychiatric Association and the ICD-11 (8A05.00) (2019) from the World Health Organisation.

Compulsivity and Tourette syndrome

According to the Diagnostic and Statistical Manual of Mental Disorders (5th edition) or DSM-5, the Tourette's Disorder diagnosis entails four elements:

a   Both multiple motor and one or more vocal tics have been present at some time during the illness, although not necessarily concurrently.
b   The tics may wax and wane in frequency but have persisted for more than 1 year since first tic onset.
c   Onset is before age 18 years.
d   The disturbance is not attributable to the physiological effects of a substance (e.g. cocaine) or another medical condition (e.g. Huntington's disease, postviral encephalitis).

It was not *her*, it was her *brain*. She was told that her brain was wired differently, therefore, it sent the wrong signals to her body. She was deemed to have a neurodevelopmental disorder[2]. Her brain malfunctioned, and with medication and behavioural therapy, the interactions would become more 'controllable'. As such, the interactions had become rendered a structural problem of her brain, and she did not need to make sense of every individual 'symptom', as what they had come to be known as. It was all in her head, and it was her nervous system that needed to be fixed. Whatever it was that made her do these things, it was not *her* fault: it was a biological issue, not an existential one.

In neuropsychiatric study of compulsivity in the diagnostic context of Tourette syndrome (or Tourette's in short), compulsive acts have been conceptualised as 'compulsions' (e.g. Shapiro et al. 1988, Shapiro and Shapiro 1992, Robertson and Cavanna 2007). 'obsessive/compulsive symptoms' (e.g. Eapen et al. 1994, Cavanna et al. 2009), 'repetitive behaviour' (e.g. Leckman et al. 1994, Miguel et al. 2000, Neal and Cavanna 2013), 'repetitive phenomena' (e.g. Cath et al. 2001), 'complex motor tics' (e.g. Verdellen et al. 2008), and 'compulsive-like tics' (e.g. Robertson et al. 2008). This makes the performance and associated sensitivities of these acts part of the 'tic' category. Tics are described as "sudden, rapid, recurrent, nonrhythmic motor movements or vocalizations" (DSM-5, 2013: p. 81). The compulsions that are the focus of this book are those that are physically interactive and involve unwanted touching, ordering and creating symmetry and balance (see Cath et al. 2001, Worbe et al. 2010).

The diagnosis of Tourette syndrome and the neuropsychiatric, biomedical and clinical histories of tics therefore provide a 'cultural' scientific context in which compulsive interactions have been understood and studied. Indeed, echoing Howard Kushner (2008: 552), "[a]lthough the DSM is a diagnostic tool, its Tourette's syndrome typology frames research protocols, restricting data to that which fulfil the criteria". What we know of tic-like compulsivity is thus the product of the evolving structures that produce particular knowledge aiming at achieving particular goals and that are underpinned by particular pre-occupations. Olivier Walusinski's book (2019) and Howard Kushner's book (1999) provide in-depth descriptions of the elaborate histories of the condition; demonstrating the strikingly broad range of explanations for tics that have proliferated over the past 140 years (see Chapter 2).

In Tourette syndrome research, interactive dimensions have not been considered to be of special interest in studying tics, and complex compulsions have remained a fringe phenomenon for their difficulty to capture by the positivist research traditions of neuropsychiatric, biomedical, and clinical sciences employed in tic studies (see Chapter 2). This has often rendered them either absent from, or indistinctively present in such research, and their conceptualisation is conflated with the entire symptom category of tics. This conflation is primarily due to the shared premise of sensibilities that surround them. Indeed, resembling tics, such as eye-blinking, shoulder-shrugging, and nose-twitching, interactive compulsions are performed either without any noticeable incentive or preceded by urges that feel like bodily sensations and unqualified anxiety that become increasingly uncomfortable with time passing when not acted upon (see Leckman et al. 1993, Banaschewski et al. 2003, Woods et al. 2005, Capriotti et al. 2013). This differs from those compulsions that are aimed at "neutralizing obsessions" that are often experienced as emotionally charged and driven by fear, as is associated with compulsivity in Obsessive Compulsive Disorder (OCD) (DSM-5 2013: 241). Nonetheless, tic-like compulsions share obsession-related compulsions' repetitiveness and the overwhelming need to do them according to strict rules. Both kinds of compulsions can be rather time-consuming and are never done for pleasure. Rather, they help address a sense of something being uneasy, not right, or incomplete, and the act is resolved when those sensations are lifted, which can lead to people avoiding certain activities, places, and people (DSM-5, 2013).

Furthermore, according to the DSM-5 compulsions "are not connected in a *realistic* way with what they are designed to neutralise or prevent, or are *clearly excessive*" (page 237, my emphasis). Their apparent unrealistic and excessive denotation is precisely where the normative dimension locates and forms the basis for the problematisation of compulsivity. As this is precisely the point where nuance is lacking and further examination is needed. Yet, these are the best approximations available of the phenomena as understandable from a diagnostician's informed outsider perspective to name and categorise bodily action like Laura's interactions. In no way do these clarify what is happening with the nuance that Laura and others with compulsive sensibilities would benefit from. However, despite having distinctive spatial connotations, there is virtually no knowledge about the circumstances under which compulsions take places in the Tourette syndrome context.

The neglect of the circumstances of compulsivity in research can largely be attributed to 'models' or 'cases' that form the context for theorisation in the neuropsychiatric sciences. Conceptualising compulsive interactions exclusively in terms of the brain and nervous system requires a detectible difference in structures, fluids, and processes from non-compulsive 'normal' brains. This requirement for a biological difference renders compulsivity an epistemological constant, which assumes compulsive bodily performance expressions of an underlying neuropsychiatric problem (see Davies 2016, Pykett 2017). This renders an ontological dispersal of Tourette's beyond the brain, or body, obsolete. Therefore, looking at what the body actually does, when it does it, and how, therefore becomes entirely superfluous. However, such an understanding cannot explain why certain compulsive engagements always happen in particular situations, with particular body parts, with particular objects or organisms, in particular spaces.

The argument in this book does not challenge the biomedical and clinical understanding and treatment as such, because it delivers on the goals it has identified as worthy of pursuing as it addresses the problems it has formulated. What the book offers is a re-evaluation of some of the principles on which both these problems are based, and in line with them, offers new directions for the development of support structures. Shifting the ontological focus from the nervous system, biological body, and person to the compulsive act then provides a new idea as to the constitution of those compulsions that happen to be in interaction with the environment. It allows for carving out a new collection of compulsive acts that is new in the sense of its regrouping of the existing categories, and developing new means of understanding them through their relations with the bodily environment. Therefore, a context-sensitive geographical approach to compulsive performativity and subjectivity can provide an explanation and even a counter argument as to why compulsions are performed as they are. This offers new avenues for a better understanding of the lives of people who have to spend a lot of their energy and time performing compulsions, resisting the urges, and/or having had to find explanations for their movements.

This onto-epistemological shift also serves as a warning for the dangers that present themselves when bodily actions that are not quite understood within rationalised scientific frameworks are problematised and medicalised. Indeed, compulsivity has been problematised against the liberal humanist depiction of humanity as moral, virtuous, and capable of understanding and rationalising its behaviour, and being in full control of its action. This has been played up in the human sciences at the risk of neglecting of dimensions of humanity outside that. Rendering it abnormal effectively halts the ethical necessity to accept the human as compulsive. Compulsivity thus provides a glimpse of what the world looks like outside moral, rational, emotional, and virtuous dimensions. Exploring this glimpse opens up the possibility to assess how human interaction with the non-human works beyond rationalisation and meaning on human terms.

Posthumanist compulsivity: A geography

In focussing on elements of bodily performance that are not derivative of signification and rationality and relying on frameworks that analytically centre the human, posthumanist approaches are helpful. In focussing instead on the material, and temporal circumstances of the acts, the relations between the body and its surroundings become central. For context, posthumanism has been attractive in geographical questions about the spatiality of the subject that seek to expand beyond postmodernist and feminist identity concerns, and invocations of universal meaning formation in humanist work. In challenging the body as a 'recipient' of spatial forces, its performativity has become rendered actively productive of spatial knowledge (see e.g. Thrift 1996, Hayles 1999, Castree et al., 2004, Castree and Nash 2006, Coyle 2006, 2007, Lorimer 2009, Braidotti 2013, Anderson 2014). Such notions place emphasis on the unfolding of the *momentary* and *in situ* performance of spatial relations with which the subject emerges, which gives rise to more vitalist considerations of bodily performativity at the expense of the conviction that humans are entirely in charge of their bodily movements. This repositions the body as intimately entangled with its constituencies (Anderson and Harrison 2010). Bodily performativity, such as compulsive acts, then allows exploring these entanglements, but from a less anthropocentric stance.

As the ontological shift from the human to the relations between the human and non-human requires an equalisation of their capacities, a focus on bodily performativity, which instigates this, requires a deprioritisation of sense-making processes, rationality, normativity, and meaning. The co-constitution of human performativity with non-humans such as objects and spaces then challenges the subject-object divide in favour of subjectification processes (see Lea 2009, Ash and Simpson 2016). Therefore, a conception of compulsivity as constituted in a 'distributed' sense calls for a hesitation in understanding body–world interactions as always and necessarily governable by the human. Whilst this further undermines the ontological primacy of the human in non-representational theories, compulsivity can help rethink the possibilities for human agency (Beljaars 2020).

Staying with compulsivity as reality that is as deserving of rigorous analysis as other, non-pathologised realities necessitates calling into question the spatial production of meanings as driving human action, as well as the absolute necessity of meaning to always underpin human action all together. This is not to challenge the humanistic and critical progress made in understanding human action, nor is it a renewed call for a 'biologisation' of human action. Rather, it opens up for envisioning a new kind of understanding of human spatial action that demonstrates what resonances between bodies and surroundings articulate. In addition to demonstrating what is beyond meaning, compulsions thus remark on changes in, and productions of, embodiment. Through the 'micro revolutions' that these entail – in the sense that a revolution signals a complete break with the past – compulsivity manifests an incessant return to the *here and now*, and, as such, reveals certain circumstances under which meaning takes hold.

Compulsivity is not the only medicalised bodily performance that has seen new understandings emerge from a posthumanist, (neo)vitalist analysis over the past two decades. Medicalised aspects of bodily performativity have traditionally been understood as an exclusively human affair in which the environment is articulated as a barrier (e.g. Gleeson 1999). However, following posthumanistic lines of inquiry, (dis)ability, (ill)health, and impairment have been reconsidered as co-constituted beyond the human (e.g.Hansen and Philo 2007, Bissell 2009, Macpherson 2010, Colls 2012, Andrews et al. 2014, Duff 2014, Andrews 2017, Gorman 2017, Hall and Wilton 2017). In accordance with such (neo)vitalist ideas, these posthumanist analyses then produce compulsivity as a *more-than-human condition*, pervading a "corporeally excessive understanding of medical conditions" (Beljaars 2020: 292). As this unsettles the normative boundaries of the pathologised realities, the societal structures on which such the principles for pathological othering are based equally become unsettled (also see McPhie 2019).

With this book coming to fruition in an age of rising fascism in the West, stringent legal systems, and increased surveillance, it also hopes to help counter the tendency to use neuropsychiatric knowledge, and the diagnostic system on which it is based, as unproblematic method of policing people. The chapters to follow put compulsivity into the context of the quest for answers as to how humanity has put itself in its current immoral situation with regards to the climate crisis. A short answer being the adoption of capitalism as the globally dominant socio-economic system; the longer one being the conditions under which choices are made under capitalism. Compulsion should be considered as playing a part in these conditions, because it does not only explain our current view of the human (as not compulsive at all) and the ousting of the phenomenon from Western societies through medicalisation, and in the rationalising sense-making exercises in everyday situations. The book contributes to finding a way out of this by arguing that it is the conjunction of the body (which includes the brain) with the bodily environment that makes compulsion possible. The understanding of compulsion as the spatially dispersed phenomenon developed in this book then becomes a mildly anti-capitalist endeavour, which bears similarities to Deleuze and Guattari's schizoanalysis (2004 [1972]).

In conjunction with a non-exhaustive engagement with the medical and clinical sciences on their understanding and theorisation of compulsive acts, the posthumanist geographical analysis to follow has four new elements of understanding on offer: first, it offers a broader conceptualisation of what counts as the 'environment' of compulsions. Second, it offers a method that is rooted in posthumanist philosophy that has a sensitivity towards different kinds of compulsive performances (i.e. bring what actually happens into the analysis). Third, it offers a constructive method of engagement with experiential accounts of the Tourette's condition more broadly, and on the basis of that, fourthly, it offers new pathways to improve the efficiency of a range of existing treatment options, but lays the foundation for a new generation of treatment).

Capturing compulsive geographies: A qualitative study

The study on which the book is based (Beljaars 2018) asks how compulsive interactions, performed by people with a Tourette syndrome diagnosis, could be affected by the bodily environment. It aims to uncover what aspects of the bodily environment play a role in the constitution, unfolding, and affective legacies of compulsions. Furthermore, it asks how the bodily surroundings became part of the negotiation of compulsivity, for instance in anticipating being compelled to act compulsively, and through offering disguise, and enabling suppressive efforts. The study put these questions in the context of the relations between people and place and their effects on personal well-being. It pursued answers to how the notion of compulsion, as medicalised performance would elucidate new understanding of the body and the socio-material environment; as such, but also in light of human thriving and floundering. Finally, the study examines how a geographical perspective might advance insights into the nature of the suffering of people who are affected by compulsivity and have a Tourette syndrome diagnosis, and what calling for new audiences to their voices may incite.

For eight months in 2015, I collaborated with 15 people who perform compulsions and wanted to consider them in more depth with me as a knowledgeable and non-judgemental stranger. The seven participants with feminine pseudonyms were femme-presenting and the eight participants with a masculine pseudonym were masc-presenting. Unfortunately, the recruitment strategy was unsuccessful in recruiting Black and Brown people, and the study only entails white participants. The participants decided to be part of the research after having been made aware of it through social media calls, targeted approaches and professional recommendation within patient and clinical networks in the Netherlands. One participant came forward on their own volition. All had a Tourette syndrome diagnosis, but the vast majority also had other diagnoses, such as OCD, Attention Deficit Hyperactivity Disorder (ADHD), autism, depression, and anxieties. The study was therefore created and conducted in the clinical and social context of the Tourette syndrome diagnosis, but did not adopt the structures it imposes on the phenomenon. Also, focussing on the interactions prevented the Tourette's diagnosis to be mobilised as a self-referential explanation. In line with this, participants are not referred to on their capacity as 'patients', as the study did not take place in a medical or clinical context. To ensure the highest level of nuance, all aspects of the study, including written and oral information was provided in Dutch, which is their and my own native language. All participants provided informed consent, agreed to have their data treated as confidential, and were offered, and accepted to appear under a pseudonym that reflects their gender.

The capturing of the spatiality of compulsive interactions took place in three ways as each method offered documentation of different aspects. This included semi-structured interviews, participant-observations, and mobile eye-tracking. The participants decided how, where, when and for how long they took place, as I left it up to them to decide with what situations they felt comfortable in discussing and performing their compulsions. Most places they chose held a deep familiarity and included homes, cars, trains, a bus, supermarkets and shops, natural areas and streets in residential areas, pharmacies, a university building, and a café. Based on semi-structured interviews, the experience of doing, as well as the sense-making processes that surround having to do individual compulsive acts became visible amidst the chaotic unfolding of everyday activities. Experiences of compulsions captured association with particular objects and spaces, how they were carried out in particular ways or rendered invisible, how prior knowledge of a space mattered, how habits interjected, and how treatment interventions mediated different situations. The interviews also provided new insights into different views of compulsions; what it said about themselves, their body and its capabilities, normativity and social othering, as well as their place in Dutch society.

These interviews proved challenging in the sense that participants found it rather difficult to discuss these acts over which they felt minimal to no ownership, which often left them searching for words that would adequately capture the experience. Many participants demonstrated some compulsions, which included touching, compulsive organisation of objects, or entire rooms that had been brought in line with compulsive requirements; presenting spaces of the house as an 'archaeology of compulsive knowledge'. Occasionally, evocative compulsive situations arose, such as when 'Elisa' received a parcel during the interview. After she closed the front door and stepped back into the living room where I am sat, she says:

> This one is for my partner, but I always <u>have</u> to unpack parcels *rips plastic* so I just can't leave it for him to do it. Parcels need unpacking *opens box* so this is <u>pure</u> compulsivity you're witnessing! *grins*

The study thus renders visible the ways in which, and the extent to which a person is involved in compulsions. Beyond this specific problem of communicating qualities of compulsion, some acts might not be spectacular or disruptive enough to leave a memory, and sink in the body-memory in a similar fashion as some habits do (see Bissell 2011). Moreover, in the communication of pathologised experiences that are not necessarily shared across humanity (Davidson and Henderson 2010), and in more general terms, in the communication of experiences, elements always get lost according to Giorgio Agamben (1999). These elements are, however, therefore not unimportant. Rather, these nonrelational aspects of experience tell us something about human-world relations that are visceral, fleshy and deal with substance (Deleuze and Parnet 1987, Harrison 2007, Anderson and Harrison 2010, MacPherson 2010).

The participant-observations centred bodily dispositions and treated movements and sensations as 'intelligence-as-act' (Melrose 1994, Dewsbury 2009: 327). These observations captured what body parts, objects, and spaces directly became involved in the compulsion. They also demonstrated how these interactions took place in real-time in between other-than-compulsive life, how the presence of other people was negotiated, and if and how they camouflaged performing them in any way. In analytically prioritising compulsive life and not invoking a dichotomy, I employ the term 'other-than-compulsive' – rather than the descriptor 'more-than-compulsive' – life. This allows accounting for dimensions of life beyond – not outside – the various stages of individual compulsions, to place them on equal footing, and to allow for the epistemological presence of both simultaneously. The observational method most clearly demonstrated how compulsive processes unfolded differently between people. Whilst many observed compulsions did not involve actual participation on my part, the method did allow approaching the affective registers of compulsive performances as close as possible through my own body (after Crang 2003, 2005, Longhurst et al. 2008, Macpherson 2010).

Observation activities were chosen in terms of the increased likelihood for evoking compulsive interactions, and had me walking seven dogs for 2 hours, felting crouched on the floor, hanging out in a charity shop, and sharing in lunch with their partners and families. The sessions lasted between 20 minutes and 3 hours, which depended on the activity and the energy and concentration level of the participant at that time. The amount of compulsions I could identify differed greatly amongst participants, ranging between three and 63 per session, and varying greatly in complexity and kind. Furthermore, retaining a friendly and informal communication style, I kept our conversations going during the observation (as recommended by Miller et al. 1998), which helped the participants to remain as comfortable as possible whilst they were being observed. Also, I used multiple strategies to be as discrete as possible in observing the participants and being in public with them. Based on the observations and extra video material I was given by some participants, it became clear how elusive these interactions are[3], and that participants were not always sure about the compulsivity of particular bodily engagements.

The mobile eye-tracking method added a first-person visual and auditory sensory perspective on compulsive acts. This method is as yet uncommon in qualitative social scientific research and had not been used for Tourette's research at the time. As it captured the unfolding of compulsive interactions in real-time and from the first-person perspective, it provided invaluable information about the circumstances that led up to them. I used the Tobii Eye-Tracking Glasses 2 system, which was one of the newest in 2015. The system consists of a pair of glasses that traces the gaze of the wearer and records it against what they are looking at, a battery and storage. The data it collected entailed a video with a red dot on it that marked the gaze location. However, as in the vast majority the calibration of the glasses to the eyes of the participant did not work, the gaze location did not play a role in the analysis of the footage. In combination with the interviews on the basis of the recordings that allowed participants to recall the moment's sensations and appeal, and decision not to suppress it, this method opened up new dimensions of access to the compulsive situation.

The mobile eye-tracking sessions would record an activity chosen by the participants that would evoke compulsive interactions. Depending on the activity and the potential anxieties surrounding the method, the recording sessions would take between two and ninety minutes. Activities such as cleaning, tidying up, gardening, performing bedtime rituals, and other typical household chores would take up to 10 minutes, after which an audio-recorded discussion would take place on the basis of the video footage if the participant wanted to. This gave us the opportunity to consider what movements were compulsive, and what made the result 'just-right enough' to continue with other-than-compulsive activities. Other activities took longer, and included grocery shopping, driving, going for a walk, which gave good insights into the different environments from the home and the different kinds of compulsions. In turn, making and having lunch, and drawing demonstrated the various levels of detail in compulsive perceptions and concurrent engagements. The method does produce privacy issues, in terms of the recorded faces, spaces, and objects, and potentially embarrassing gaze fixations. Therefore, during the sessions, I closely registering any anxieties, providing distraction and assurance, and did not to watch the live-stream when it helped to reduce anxieties and discomfort. The majority of participants who did partake in this method were curious to see their compulsions, whilst others did not have the time or did not want to watch the footage after the recording session(s).

The analysis of all data was not based on clinical categories, but rather on the *accomplishments* of compulsions: for instance, on specific tactile, auditory, visual sensations, new orders of objects in accordance with each other and/or spaces, and feeling just-right again, etc. The analysis of all data took place in Dutch to preserve the thrust and emotional affections of the statement and the personality of the participant. Languages are ontological systems, and the English language and the way it is used shapes a particular ontology that differs slightly from the Dutch language ontology. Some truths, systems of relations, and expression of experience do not translate with adequate precision. Therefore, I formulate the translations in a way that expresses the meaning in English at sentence-level that priori-tises readability in English but diverts slightly from the Dutch words in the transcripts. These are minute differences, but differences nonetheless. The interviews were fully transcribed, thematically coded with a small set of codes that was used for all analysis in NVIVO-10, and translated to English where quoted in the book. The observation data was detailed in descrip-tion in Excel and coded in NVIVO-10. I watched most of the eye-tracking videos back at 50% of the speed, as some compulsions were very brief, in Adobe Premiere Pro CC 2015. The interviews on the basis of the recordings were transcribed and coded where relevant and also the notes from my field diary were digitised and coded.

It is important to note here that compulsivity is not derivative of men-tal health nor mental illness, and compulsions are thus not conceived of as a priori signs of (dis)ability, (ill)health, impairment, or as affirmation of abnormality. In the chapters that follow, it does not adhere to prin-ciples of disease, disorder, defect, or any kind of weakness. Laura and many others I have spoken with about their compulsive interactions refuse to see themselves necessarily as victims of their own condition (see also Hollenbeck 2003). In fact, the theory developed in this book suggests quite the opposite; it testifies of people who have developed highly sophisticated mediations of their bodily involvement in compulsive interactions. A more empathetic and less normatively prescribed understanding invokes com-pulsivity as a particular kind of body-world formation that more easily sways some bodies; those diagnosable with Tourette syndrome in particu-lar. Therefore, although the book is never in denial, and remains in full awareness, of compulsivity's potential to burden, it does not aim to 'fix' the afflicted body, or extend a search for a 'cure'. Rather, with the book I aim to open up a space in which those affected can express the "affective force" of their experiences, as Felicity Callard (2006: 875) puts it. Instead of necessarily translating their experiences into medical language, the book hopes to contribute to the formalisation of an experiential vocabu-lary that captures the richness of individual sense-making processes. To this end, I hope that the book facilitates a discussion that is heard beyond their community and carries to other realms that can be of assistance in living and coping with compulsive interactions, and the Tourette's condi-tion more broadly.

Outline of the book

This book does three things. First, it provides a detailed description of the everyday life of people with Tourette syndrome, and establishes the com-plexities of the inescapable situatedness of their compulsions, with which I problematise the dominance of the universalist and model-based approaches that currently govern the life sciences that study Tourette syndrome and the asymmetries in knowledge construction this purports. Second, it develops compulsivity as a spatially dispersed phenomenon that demonstrates how embodied life is incessantly caught up with its material environments. Compulsion is then explored as a human action that is instigated beyond the human, and folded back onto the human. And third, it posits compulsion as problematic phenomenon in academia because of its anti-humanist, a-moralist, and irrational connotations, because it demonstrates a side of the human we do not like and that is problematised and medicalised in Western societies.

Chapter 1 traces the confusing sense-making exercises that people with Tourette syndrome go through when confronted with their compulsions. Considering the psychogenic theories that have been denounced, allows exploring the relations between the acts and the self and psyche. Compulsions are found not to adhere to these introspective constructs. Building on the absence of meaning, the chapter follows participants' querying of compulsions as holding wisdoms or practical usefulness. As such ideas also do not hold up, the remaining explanatory gaps are considered, which leaves us with the conclusion that meaning, rationality, and purpose are dead ends in formulating explanations as to why compulsive interactions take place.

The confusions live on as the neuropsychiatric sciences cannot answer all questions individuals with compulsive sensibilities may have. Chapter 2 outlines that current knowledge of Tourette's-related compulsions is minimal, which, as a medicalised phenomenon, is mainly a result of the onto-epistemological structures that govern the neurosciences, psychiatry and psychology that study Tourette syndrome. It elaborates on how these structures create an impasse in the study of compulsions *as such*, but also as connected to the bodily surroundings. Furthermore, it problematises the current limited involvement in research of people with Tourette's who perform and experience the various circumstances of compulsions. Based on this critical review of these life sciences the argument in Chapter 2 identifies four transformations people's understanding of compulsivity goes through, and sets out how a de-problematisation of the bodily action helps to expand research horizons.

Chapter 3 unpacks what analytical directions are needed to understand compulsivity better as situated phenomenon. It elaborates on the infinite variety with which individual compulsions take place, discerns how people with Tourette's distinguish compulsivity from other movements and activities, and troubles the clear conceptual delineation between these practices. Tracing perception and how awareness and memory come into play, the chapter also establishes how they are performed during other-than-compulsive daily life activities, or how they disrupt or even aid such activities.

Chapter 4 explores the bodily and sensory processes leading up to compulsions; particularly regarding the increasing urge to act compulsively. Developing how compulsivity could be regarded as emerging from, and responding to a 'landscape of tension', it illumines the manifold ways in ways people with Tourette's feel drawn into performing compulsive acts and how they make sense of this urgency. This includes a spatial knowledge that presents itself as (not-)just-rightness, and as recurring understanding of the urge and compulsion combination as 'energy' that is produced or reproduced in, or needs release from, their bodies. The chapter establishes how these experiences of becoming compulsive and the 'achievements' of individual compulsions suggest a particular, new, set of relations between the body and its environment.

Chapter 5 takes the suggestions of the previous chapter forward, by establishing the ways in which the body is involved in individual compulsions and the urge to act compulsively. As the body feels out of control of the affected person, it analyses the role of the body prior to, during and after a compulsion in the relations between the body and its surroundings. Its fleshiness and its spatial position in relation to the bodily surroundings become important in compulsions, which can be understood as 'configurational' process. The chapter does so by considering the bodily strategies with which people with Tourette's negotiate being in a particular place or being confronted with a particular object. Furthermore, it considers how current behavioural and pharmacological treatment affects the relations between the body and its surroundings. Based on this, the chapter sets out a renewed understanding of the body and its performativity as subservient to the socio-material environment.

Building on the previous chapters, Chapter 6 discusses how the bodily surroundings come to appear prior to, during and after compulsive interactions. It traces the ways in which objects and spaces become powerful through their different capacities to affect a body, and it considers how the vitality of objects and spaces are negotiated and neutralised. The extracorporeal configurational theory suggests how a renewed understanding of objects and spaces is can be formulated. This chapter ends by developing the foundations for the spatial theory of compulsion. This theory renders compulsivity not as a (mal)functioning of the brain or body, but as an expression of the body *and* its spatial situation. As such, it denotes the becoming- and un-becoming-compulsive of the situated body by means of a dynamic configuration of bodies, objects and spaces.

Chapter 7 builds on the spatial theory of compulsion developed in Chapter 6. It does so by exploring how and when the 'body-object-space configurations' that lead to compulsions become important in the life of people with Tourette's. The chapter continues by explaining how these configurations are negotiated through various kinds of 'stabilisations' of the body and its surroundings to avoid compulsive engagement. As the duration of the stability of these configurations differs, a spatial ecology emerges of objects and room features that need re-stabilisation under different circumstances. Considering such strategies and the consequences for the wellbeing of the person involved, this chapter lays out a compulsive personal geography of everyday places.

Chapter 8 continues on the compulsive ecological production of spaces by presenting how people deal with the stabilisation efforts through, for instance, following routines, developing habits, (re)designing rooms, doing activities in specific ways, and by managing other people's presence. It then considers the possibility for such efforts in, for instance, public places and different social situations to reduce anxieties and uphold personal wellbeing. Such mediations cave out new ways of attending to various aspects of the body in its relations to the bodily environment and its organisations. To this end, the chapter puts forward how these findings inform new directions for support for people with Tourette syndrome as well as new kinds of future research priorities.

Chapter 9 considers how the spatial theory of compulsivity and the novel understanding of a dispersed sense of wellbeing could be of interest to a broader humanity beyond people with Tourette's. As such, it outlines how compulsivity can be understood to underpin human spatiality. It discusses, therefore, what it means to understand humanity as inherently compulsive, and what a situated understanding of compulsion does to our understanding of humanity as conceptualised in liberal humanism by Western societies. Additionally, the chapter thinks through the potential consequences for the ways in which we understand human beings as having capacities to act compulsively in addition to rationally and emotionally. It argues that the politico-ethical implications are especially significant in relation to capitalism and human exceptionalism, and in challenging normativities surrounding bodily action, it suggests new principles for an empathetic posthumanity.

Notes

1 Laura permits the usage of her name and narration of some of her compulsive experiences for the purpose of this monograph.
2 At the time, Tourette syndrome was conceptualised as a neuropsychiatric disorder, whilst in 2020, it is understood as neurodevelopmental disorder. Neurodevelopmental disorders are deemed structural problems with the nervous system that is not present at birth and manifests with the development of the young brain. Chapter 1 discusses this further.
3 In recognising compulsive from other-than-compulsive interactions my sensibilities developed through having lived with my diagnosed sister, as well as through extensive discussions with her and Dr Cara Verdellen. 'Learning' to recognise compulsive engagement per participant was helped by all prior meetings (including with recruitment attempts), which gave me an idea what to look for during the observations. Before and during the observations, I looked for changes in the smoothness and disruption of movements, as well as

the way in which movements fitted with their intentional engagements. In case I felt that these did not fit with these acts I would write them down as compulsions, and check with the participant as and when this was possible. In some instances, such as with Mina, participants 'taught' me how to do a particular touch tic by telling me synesthetic intricacies of their engagement. As such, on occasions, I joined their embodiment without claiming to feel what they felt, but to grasp the felt knowledges that pervade compulsive interactions (after Haraway 1991). In doing so, I rendered my own body 'responsible' to our shared environment, and could be understood as a way of being sensitive on terms of another (Mol, 2002, Despret, 2013).

References

Agamben, G. 1999. *Remnants of Auschwitz*. In: McCormack, D.P. ed. *An Event of Geographical Ethics in Spaces of Affect. Transactions of the Institute of British Geographers* 28, pp. 488–507.

Anderson, B. 2014. *Encountering Affect: Capacities, Apparatuses, Conditions*. Farnham: Ashgate.

Anderson, B. and Harrison, P. eds. 2010. *Taking-Place: Non-Representational Theories and Geography*. Farnham: Ashgate.

Andrews, G.J., Chen, D. and Myers, S. 2014. The 'taking place' of health and wellbeing: Towards non-representational theory. *Social Science and Medicine* 108, pp. 210–222.

Andrews, G.J. 2017. *Non-Representational Theory and Health: The Health in Life in Space-Time Revealing*. London: Ashgate/Routledge.

Ash, J. and Simpson, P. 2016. Geography and post-phenomenology. *Progress in Human Geography* 40(1), pp. 48–66.

Banaschewski, T., Woerner, W. and Rothenberger, A. 2003. Premonitory sensory phenomena and suppressibility of tics in Tourette syndrome: developmental aspects in children and adolescents. *Developmental Medicine and Child Neurology* 45, pp. 700–703.

Beljaars, D. 2018. Geographies of Compulsive Bodies: Bodies, objects, spaces. PhD dissertation, Cardiff University.

Beljaars, D. 2020. Towards compulsive geographies. *Transactions of the Institute of British Geographers* 45, pp. 284–298.

Bissell, D. 2009. Obdurate pains, transient intensities: affect and the chronically pained body. *Environment and Planning A* 41(4), pp. 911–928.

Bissell, D. 2011. Thinking habits for uncertain objects. *Environment and Planning A: Economy and Space* 43, pp. 2649–2665.

Braidotti, R. 2013. *The Posthuman*. Cambridge: Polity.

Callard, F. 2006. The sensation of infinite vastness; or, the emergence of agoraphobia in the late 19th century. *Environment and Planning D: Society and Space* 24, pp. 873–889.

Capriotti, M.R. et al. 2013. Environmental factors as potential determinants of premonitory urge severity in youth with Tourette syndrome. *Journal of Obsessive-Compulsive and Related Disorders* 2, pp. 37–42.

Castree, N., Nash, C., Badmington, N., Braun, B., Murdoch, J. and Whatmore, S. 2004. 'Mapping posthumanism: An Exchange' (2004). *Environment and Planning A: Economy and Space* 36(8), pp. 1341–1363.

Castree, N. and Nash, C. 2006. Posthuman geographies. *Social & Cultural Geography* 7(4), pp. 501–504.

Cath, D.C. et al. 2001. Repetitive behaviors in Tourette's syndrome and OCD with and without tics: what are the differences? *Psychiatry Research* 101, pp. 171–185.

Cavanna, A.E. Servo, S., Monaco, F. and Robertson, M.M. 2009. The behavioural spectrum of Gilles de la Tourette syndrome. *Journal of Neuropsychiatry and Clinical Neuroscience* 21(1), pp. 13–23.

Colls, R. 2012. Feminism, bodily difference and non-representational geographies. *Transactions of the Institute of British Geographers* 37, pp. 430–445.

Coyle, F. 2006. Posthuman geographies? Biotechnology, nature and the demise of the autonomous human subject. *Social and Cultural Geography* 7, pp. 505–523.

Crang, M. 2003. Qualitative methods: touchy feely look see? *Progress in Human Geography* 27, pp. 494–504.

Crang, M. 2005. Qualitative methods: there is nothing outside the text? *Progress in Human Geography* 29(2), pp. 225–233.

Davidson, J. and Henderson, V.L. 2010. Travel in parallel with us for a while': Sensory geographies of autism. *The Canadian Geographer/Le Geographe Canadien* 54(4), pp. 462–475.

Davies, G.F. 2016. Remapping the brain: Towards a spatial epistemology of the neuro-sciences. *Area* 48, pp. 125–125.

Deleuze, G. and Guattari, F. 2004 [1972]. *Anti-Oedipus: Capitalism and Schizophrenia*, trans. Massumi, B., London: Bloomsbury.

Deleuze, G. and Parnet, C. 1987. *Dialogues*. New York: Columbia University Press

Despret, V. 2013. Responding bodies and partial affinities in Human–Animal worlds. *Theory, Culture and Society* 307(8), pp. 51–76.

Dewsbury, J.-D. 2009. Performative, non-representational, and affect-based research: seven injunctions. In: DeLyser, D., Herbert, S., Aitken, S.C., Crang, M.A. and McDowell, L. eds. *The SAGE Handbook of Qualitative Geography*. London: SAGE, pp. 321–334.

Duff, C. 2014. *Assemblages of Health: Deleuze's Empiricism and the Ethology of Life*. Dordrecht: Springer.

Eapen, V., Mortiarty, J. and Robertson, M.M. 1994. Stimulus induced behaviours in Tourette's syndrome. *Journal of Neurology, Neurosurgery, and Psychiatry* 57, pp. 853–855.

Gleeson, B. 1999. *Geographies of Disability*. London and New York: Routledge.

Gorman, R. 2017. Therapeutic landscapes and non-human animals: the roles and contested positions of animals within care farming assemblages. *Social and Cultural Geography* 18(3), pp. 315–335.

Hall, E. and Wilton, R. 2017. Towards a relational geography of disability. *Progress in Human Geography* 41, pp. 727–744.

Hansen, N. and Philo, C. 2007. The normalcy of doing things differently: bodies, spaces and disability geography. *Tijdschrift voor economische en sociale geografie* 98(4), pp. 493–506.

Haraway, D. 1991. *Simians, Cyborgs and Women: The Reinvention of Nature*. London and New York: Routledge.

Harrison, P. 2007. 'How shall I say it?' Relating the nonrelational. *Environment and Planning A* 39. pp. 590–608.

Hayles, N.K. 1999. *How We Became Posthuman: Virtual Bodies in Cybernetics, Literature, and Informatics*. Chicago, IL: The University of Chicago Press.

Hollenbeck, P.J. 2003. A Jangling journey: life with Tourette syndrome. *Cerebrum* 5(3), pp. 47–61.

Kushner, H.I. 1999. *A Cursing Brain? The Histories of Tourette Syndrome*. Cambridge, MA: Harvard University Press.

Kushner, H.I. 2008. History as a medical tool. *Lancet (London, England)* 371(9612), pp. 552–553.

Lea, J. 2009. Post-phenomenological geographies. In: Kitchen, R. and Thrift, N. eds. *International Encyclopaedia of Human Geography*. London: Elsevier. pp. 373–378.

Leckman, J.F., Walker, D.A. and Cohen, D.J. 1993. Premonitory urges in Tourette's syndrome. *Journal of American Psychiatry* 150, pp. 98–102.

Leckman, J.F., Walker, D.E. and Goodman, W.K. 1994. 'Just right' perceptions associated with compulsive behavior in Tourette's syndrome. *Journal of American Psychiatry* 151, pp. 675–680.

Longhurst, R., Ho, E. and Johnston, L. 2008. Using 'the body' as an 'instrument of research': kimch'i and pavlova. *Area* 40, pp. 208–217.

Lorimer, J. 2009. *Posthumanism/Posthumanistic geographies*. In: Kitchin, R. and Thrift, N. eds. *International Encyclopedia of Human Geography*. Amsterdam: Elsevier, pp. 344–354.

Macpherson, H. 2010. Non-representational approaches to body–landscape relations. *Geography Compass* 4, pp. 1–13.

McPhie, J. 2019. *Mental Health and Wellbeing in the Anthropocene. A Posthuman Inquiry*. Singapore: Palgrave Macmillan.

Melrose, S. 1994. *A Semiotics of the Dramatic Text*. London: Saint Martin's Press Inc.

Miguel, E. et al. 2000. Sensory phenomena in obsessive-compulsive disorder and Tourette's disorder. *Journal of Clinical Psychiatry* 61, pp. 150–156.

Miller, D. et al. 1998. *Shopping, Place and Identity*. London and New York: Routledge.

Mol, A. 2002. *The Body Multiple: Ontology in Medical Practice*. Durham, NC: Duke University Press.

Neal, M. and Cavanna, A.E. 2013. Not just right experiences in patients with Tourette syndrome: Complex motor tics or compulsions? *Psychiatry Research* 210, pp. 559–563.

Pykett, J. 2017. Geography and neuroscience: critical engagements with geography's 'neural turn'. *Transactions of the Institute of British Geographers* 43(2), pp. 154–169.

Robertson, M.M. and Cavanna, A.E. 2007. The Gilles de la Tourette syndrome: A principle component factor analytic study of a large pedigree. *Psychiatric Genetics* 17, pp. 143–152.

Robertson, M.M. et al. 2008. Principal components analysis of a large cohort with Tourette syndrome. *British Journal of Psychiatry* 193, pp. 31–36.

Shapiro, A.K. and Shapiro, E. 1992. Evaluation of the reported association of obsessive-compulsive symptoms or disorder with Tourette's disorder. *Comprehensive Psychiatry* 33, pp. 152–165.

Shapiro, A.K. et al. 1988. Measurement in tic disorders. In: Shapiro, A.K. et al. eds. *Gilles de la Tourette syndrome*. 2nd ed. New York: Raven Press. pp. 452–480.

Thrift, N. 1996. *Spatial Formations*. Thousand Oaks, CA: Sage.

Thrift, N. 2007. *Non-Representational Theory*. London: Routledge.

Verdellen, C.W.J. et al. 2008. Habituation of premonitory sensations during exposure and response prevention treatment in Tourette's syndrome. *Behavior Modification* 322, pp. 215–227.

Walusinski, O. 2019. *Georges Gilles de la Tourette, Beyond the Eponym*. Oxford: Oxford University Press.

Woods, D.W. et al. 2005. The premonitory urge for tics scale PUTS: Initial psychometric results and examination of the premonitory urge phenomenon in youths with tic disorders. *Journal of Developmental and Behavioral Pediatrics* 26, pp. 397–403.

Worbe, Y. et al. 2010. Repetitive behaviours in patients with Gilles de la Tourette syndrome: tics, compulsions, or both? *PLoS ONE* 5, e12959.

1 Confusions

Dead ends and (un)making sense

> There is so little of the grand or noble about my condition that, were it in my power to wave a wand and banish Tourette from humankind, I would be tempted to do so. Yet, I have to wonder whether this action would take something out of the world.
>
> (Hollenbeck 2003, A Jangling Journey: Life with Tourette Syndrome, np)

Compulsivity is confusing as phenomenon, construct, imagination, as intuitively it might come to appear as negation of well-known modes of sense-making. Does that make compulsivity the ultimate *negative* action, the feared abject gesture, the final, tragic loss of bodily control? According to those who perform compulsive acts, like Peter Hollenbeck and the vast majority of the research participants of this study, not necessarily.

Understanding compulsive acts and considering how they figure and bequeath legacies in people's lives cannot take place in sole opposition to the familiar behavioural constructs, as it would risk dismissing what Hollenbeck denotes as the 'something' that should not be taken out of the world. To come closer to a more productive notion of compulsivity than 'something', this chapter explores how compulsive interactions are made sense of in the context of everyday life and how they have a more or less complicated presence in the lives of the research participants. It does so by working through the multiple confusions that different aspects of compulsive interactions create in different dimensions of lived embodied action, and by following the participants' examinations of their compulsive tendencies and interactions through a range of 'systems of logic' that they employ to govern other-than-compulsive dimensions of their lives.

Aligned with increasing fragmentation of social life and desecularisation of spiritual life particularly in Western cultures, potentially exacerbated by ongoing connectivity with the internet, various forms of self-reflection and psychological introspection have become important coping mechanisms in fast-paced, postmodern, life. To not get lost in the chaos of everyday life or forget about the important stuff in life, activities require some level of conscious consideration, planning, and/or meaningful engagement. It is then not surprising that compulsive acts that may not lend themselves to make

DOI: 10.4324/9781003109921-2

a good fit are often still brought into structures of meaning and belief, are understood to serve some sort of purpose, and are considered along the same lines as commendable acts. Participants differ in the ways in which and to what extent they make such alignments. The compulsions for which they employ sense-making exercises also vary strongly. Following and unpacking the rationales that become applied to the compulsive interactions reveals the limits of the familiar ways of making sense of bodily action. Indeed, it becomes clear through the many contemplations and compulsive performances by these 15 people how compulsivity as a phenomenon calls the applicability of these rationales into question.

Self/psyche

"Why? I don't know" marks a typical ending to Alan's elaborations and is his go-to sentence when pointing out what object incites him to compulsively interact with. On a stroll, following the white gravelled path between the fenced off green horse pastures he often walks past, he describes why his eyes were drawn to certain elements in his surroundings; a wooden slat lying between the leaves in a forested area, windows on a farm building in the distance, a line of trees. His statement of not knowing why is not an effort to evade exploring what could be a reason for his compulsive tendencies, nor is it emotionally too difficult to convey. Over the years, he had examined his compulsions in depth, and he had concluded that there was 'quite simply' no reason. That does not mean that he suffered less from having to perform his compulsive interactions or having the sensibilities that irked him to act. It did allow him to refrain from questioning himself on every occasion.

Whilst Alan was comfortable putting this statement of inexplicability forward for the vast majority of his compulsions and tics, for other participants who discussed particular interactions, this was not possible. Over almost 20 years of trying to make sense of compulsions and tics, Tomos has considered the relations between his personality and his compulsive tendencies in detail, and is very elaborate in his musings on particular compulsive interactions. Ordering compulsions are interpreted as confirmation that he is a "neat" or "tidy" person. Such compulsive ordering acts that contribute to the basis of this assertion include putting the remote control and other small objects on the coffee table in a particular composition, clearing everything he does not use in the present moment off the kitchen counter, and spending great time and effort to reposition chairs and plant pots in the exact place they had been before he vacuums the room. Such acts may underwrite virtues such as personal hygiene, self-organisation, and self-discipline. Tomos moves between claiming them as such and considering the validity of this claim:

> It's more like, it has to be clean, it needs to be orderly … order to combat the chaos, because I don't see myself … Like friends of mine eh … leave the dishes out for weeks, eh … that I just don't understand. No, I really

can't, I really can't do that *laughs* It must be done immediately (...) 'I have to finish something', 'I must do it now and it can't wait'. That's what's really bothering me (...) if I have to wait, I get restless. No, it does play a role in my subconscious; if I do it immediately, I have some peace in doing my own stuff and doing the things I like.[1]

Tomos feels that he is compelled to compulsively 'clean' or 'finish something' and that not doing so creates problems, not only in terms of restlessness, but also in how he 'sees himself'. He elaborates:

You know, I can be really rather lazy and negligent, it's not that I do everything perfectly, definitely not, I'm not as perfect as these people who make everything spick and span all the way to the last little spot. I'd want to be like that, but at a certain level, a part of me says 'okay, this is enough'. I'm quite lax, I do have that tendency for compulsivity, but I don't have to fulfil it entirely, so I'm a little bit in the middle, and I've also kind of made it into a lifestyle to eh ... just allow it to happen, and to enjoy it sometimes as well. I can enjoy things being positioned nicely and when all of these things are put in straight lines.

Thinking through how he relates to these tendencies and how the acts figure in his life, he sees them as part of a 'lifestyle' that also says something about his character. After swallowing the last bite, he immediately washes the plate and cutlery and puts them in the cupboard. Rather than taking away fears over germs – as people who perform compulsions following on obsessions would experience – Tomos states that "I also demand that from myself, I just really want a military discipline, if I can, if that's viable ..." Some of his compulsive acts and tendencies had formed and maintained a favoured sense of self. However, when considering how he had made this work for himself and if it is a strategy that helps coping with the compulsions, he hesitates:

Well yeah, now I'm formulating it like that, I can say that, in a certain way, I have dealt with this philosophically, but saying so in hindsight, really. I have been doing these compulsions for a long time already of course.

Having examined his compulsions as a part of his life and as individual acts in depth, Tomos cannot reconcile compulsions as somehow conveying a truth about (a dimension of) himself. He, and all other participants recognise that such a truth claim does not work for all compulsions, and the acts are only attributed such truth after they are performed. However, attractive as such claims might be, they are always only made in hindsight, with the compulsion escaping an explanation in its spontaneity[2]. Indeed, if they cannot be rationalised within one's life and are beyond praise or blame, compulsions are a-personal or pre-individual, following Simondon (2020 [1964]) and Deleuze and Guattari (2004 [1980]). Herein, they are without virtue and potentially even amoral – not immoral. Action ostensibly lacking truth about the self is not only confusing, it can be terrifying. Tomos and others rather claim some compulsions as dimension of their identity, thereby averting implying they act without reason, because that would call *all* their bodily action into question.

Labouring this point with all participants, I had asked them if they thought that compulsive interactions conveyed any meaning at all. 'Alan' was the only participant who did not outright entirely dismiss the idea of compulsions being meaningful in any way, ever. He asserts "No, I don't believe that it has a meaning", but pondering the possibility of meaning he adds:

> or maybe precisely it does, but that I'm just not aware of it ... that could then potentially come up in that hypnosis session.

Whereas his examinations of his conscious understanding of his compulsive tendencies do not reveal an individualised pattern of meaning, he keeps the door open for it to emerge from his subconscious. The involvement of the subconscious as potentially holding explanatory power is not a new idea. It had been considered as frame of reference at length in psychoanalytical endeavours in the first half of the twentieth century, but has now been dismissed almost entirely in Tourette's research.

Focussing on the ostensibly rich symbolic potential of compulsive interactions, the earliest psychoanalysis of compulsions, including compulsive interactions, cursing, and other acts that had a socio-normative resonance was developed on the basis of Freud's analysis of hysteria[3] (1895). In their interpretation by the psychoanalyst, they were aligned with dimensions of the oedipal complex that Freud had developed to explain what moves people in the broadest sense of the word, and how desirous relations that underpin the social world take shape. Setting out to understand the content of tics and compulsions, psychoanalytical approaches thus focused on capturing the applied signification of acts as function of an aspect of the person.

Compulsive tendencies were considered to have partial presence in the psychic life of some of the research participants, as some acts during certain moments were still considered protruded by themselves, but they reported a sense of being made to perform certain acts by an unknown subjectivity. For instance, Sion mentioned that when he came home exhausted early in the morning after having worked the entire day and part of the night, he still had to reposition the cushions on the sofa, make sure all kids toys were put back in the play station in the living room, and he needed to clean and reorder the kitchen to its precise set-up to appease his compulsive tendencies. Sion and other participants referred to such subjectivity in terms of enslavement:

> I'm a slave of myself, it never seems to be fine (...) alas, indeed, of course I was born this way and I don't know any better

(Sion)

Tourette, precisely, how do you say that, is almost a little guy, almost a slave driver indeed because it is a slave driver because you become so anxious because of the unrest and that paranoid feeling like 'something's off'.

(Tomos)

Making sense of having to perform compulsive interactions through inciting another subjectivity has also been considered by psychoanalysts Henri Meige and Eugene Feindel. They noted a patient suffering from a split subjectivity, which fit with Freud's oedipal complex. In their 1902 publication *Les Tics et leur Traitement*[4], they personified the compulsive tendency as the child who performs the acts, and the suffering person as the father who fails to resist these acts and "remains a slave to the whims of his own creation" (c.f. Ferenczi 1921: 10). Precisely because of this conflict in the explanation of compulsion in terms of meaning, such psychoanalysis understood compulsion both as a product and as a result of a fragmented psyche. Psychotherapy that was developed and applied on the principles of this form of psychoanalysis therefore not only aimed at reducing the number of compulsive acts – and tics in the process – it also aimed at reconciling the multiple subjectivities by attempting to abolish the conditions that would cause the frictions between them. The treatment attempted this by making "improvements in the total personality" of people with a Tourette syndrome diagnosis (Mahler and Rangell 1943, c.f. Kushner 1999).

Whilst Meige and Feindel did not see compulsivity as sexuality-driven; mostly because their patients did not make reference to this dimension of Freud's psychoanalysis, Hungarian psychoanalyst and Freud's student Sandor Ferenczi did. His 1921 psychoanalysis of compulsions and tics reconceptualised compulsivity as existing primarily by virtue of sexuality. The majority of psychoanalysts that came after him subscribed to his theories. Ferenzci argued that compulsions and complex tics would be brought about by childhood (sexual) trauma and severe styles of upbringing that failed to teach a person how to express themselves, which would lead the person to resort to self-love in the form of masturbation. Whereas Wilson (1897) and Lerch (1901) argued that tics and compulsions represented masturbatory acts, for Ferenczi (1921), compulsivity "resulted from repressed masturbatory desires", where compulsive acts are released from repression as bursts of "dammed up libido". This libido was not only considered to be dammed up in the genitals, but in all organs because of "traumatic displacement of libido". He explains that "muscular actions and skin irritations carried out apparently without thought and believed to be without meaning were able to seize the whole of the genital libido" (1–2), and therefore, the whole ego. Similar to Meige and Feindel, Ferenczi also placed the analytical and therapeutic point of gravity on the performing person, not the compulsions or tics. His analysis of his patients whom he called 'tiqueurs' would consist of an observation followed by the interpretation of the compulsions of a patient (1921: 3):

A patient (an obstinate Onanist) practically never ceased to carry out certain stereotyped actions during analysis. He kept on smoothing his coat to his figure, frequently several times to the minute; in between he assured himself of the smoothness of his skin by stroking his chin or he gazed with satisfaction at his shoes which were always shining and polished. His entire mental attitude, his self-sufficiency, his affected speech couched in balanced phrases to which he was his own most delighted listener, marked him out as a narcissist contentedly in love with himself, who – impotent with women – found his most apposite method of gratification in Onanism. He came for treatment only at the request of a relative and fled from it in haste at the first difficulties.

(Ferenczi, Psycho-analytical observations on tic, 1921: 3)

Ferenczi's notes depict a person who acts compulsively for the sole purpose of gaining gratification. His psychoanalytical argument entailed that even acts that are painful and harmful to the individual should be understood as expressions of repressed self-love through the pleasure principle that governs sadism and masochism (Fenichel 1945). Performing compulsive acts to 'relieve the libidinal pressure' then presumes an anticipation of, and an upfront motivation to receive, pleasure (however painful) from the act. However, psychoanalytical accounts of compulsion claim that by virtue of the compulsive drives emerging from the subconscious, that what was gained by doing the act could not be recognised as pleasure by the afflicted person, and could only be realised to the degree that the drive for pleasure incited the person to continue performing compulsive acts (Kushner 1999). This posits compulsivity as an addiction that remains unbeknownst to the affected person, but does render them narcissistic in the process. In this regard, compulsions would signal a subconscious 'loop' that would be impossible for this person to disrupt without psychoanalytical treatment.

The assertion that compulsive acts would be driven by repressed sexuality did not come forth from the explanations and sense-making exercises by the patients: many actively challenged them. Mahler and Gross (1945), who psychoanalysed their 11-year-old patient 'Pete', saw his compulsion to touch a warm radiator as clearly symbolic of masturbatory tendency. Their psychotherapy addressing his sexual life did not 'cure' his compulsive acts, but they did register a reduction in simple tics. Pete was not convinced of the method though, as he was quoted "You told me the tics came because of the mixed-up feelings and lots of those feelings have to do with the sex business. Well, I've had tics since I was about 3 years old. How does that fit in?" (c.f. Kushner 1999: 112). In response to Pete's question, as well as mounting critique on the theory from the growing neurobiological sciences in the 1950s, Mahler argued that both a patient's willingness and complete resistance to discussing sexuality and personal sexual histories during therapy sessions could be seen as unusually strong interest in sex, which would affirm sexual repression as cause of tics (Kushner 1999).

Whereas psychoanalytical theories of compulsive acts did consider the experience of the patients under study and treatment, these were only taken as 'valid' knowledge if they fitted the psychoanalytical metaphysical framework that was based on concepts that indicated personality traits that made a claim about people's desires and state of sanity. By extension, Howard Kushner (1999) summarises the main critique of psychoanalysis of compulsivity in Tourette syndrome with the accusation of the psychoanalytical rationales to considering compulsive acts only served the extension of the psychoanalytical mandate. It did not formulate a comprehensive set of rationales to explain compulsive acts (e.g., Shapiro and Shapiro 1968).

Whilst psychoanalysis was not part of the present study, and the absence of evidence does not suggest evidence of absence, none of the research participants related performing compulsions to inciting sexual gratification, pleasure, or understood them to sustain self-love. Even with the interview and other unrecorded conversations allowing the exploration of the topic in various ways, no participant brought the compulsive aspect of their life in relation with sexuality, not even in its broadest sense. Rather, they all remarked on how compulsive tendencies and interactions distorted the connections between them as a person and their body. Although the subconscious and other subjectivities had been considered by the participants as ways of making sense of their compulsive tendencies and the interactions themselves, they were not considered answers to why compulsive interactions had to be performed. Inciting the figure of the slave driver, however, did help take pressure of having to account for some of their unwanted, unprecedented acts both for themselves and for others who are less familiar or totally unfamiliar with compulsivity, as it allowed them to shift responsibility for the unwanted engagements. These self-referential and meaningful incitations were only ever mentioned in the descriptive mode and did not fulfil an analytical function in the sense-making processes around any compulsive interactions. For clarification, the figure of the slave driver was a useful way of expressing the enormous pressure behind the compulsions, but they did not search for meaningful implications of being a slave or having a slave driver figure in their further life. Indeed, participants only incited another subjectivity in the context of having to act compulsively in general, not with reference to particular acts, body parts, nor specific compulsive tendencies or in specific places. Thus, the psyche as a specific order of understanding human action seems to insufficiently capture compulsivity.

Functionality/utility

Whilst structures of meaning in relation to the psyche were not considered helpful in understanding compulsivity, some compulsive acts were examined along functionalist lines of inquiry, hailing those with favourable outcomes. During the interview, one of Dylan's compulsive interactions demands him to touch his nose in a particular way, which coincided with dealing with a runny nose from hay fever. After performing the action several times, he considers them on their practical consequences:

> At some point I think ... this is so difficult to distinguish; is it normal or not? Am I handling my nose because it's a touch tic, or because <u>it just works</u>.

This alignment of a compulsive act with intention and a goal even leads him to question whether or not handling his nose is 'normal' or compulsive at all. If compulsive acts can serve a goal, what remains of the compulsive aspect of any act? Could compulsion theoretically be 'solved' by improving identification of the goals that these interactions seem to serve, and/or make better use of their outcomes? Would alignment of one's intentions simply annul compulsivity? In such functionalist explanations, compulsion would hold some sort of truth that could be revealed to the person during or after the act. If the person performing the compulsive interaction did not recognise its practicality, it would reflect their ignorance, rather than an absence of functionality in the act. In such explanations of compulsive acts, the body and its movements gain some sort of wisdom, but of the utilitarian kind. An ostensible inability to decipher this wisdom could then even reflect the intelligence of the person performing the act. Compulsion thus does not only highlight how other action serves a purpose that is known in advance or is savoured as meaningful, simultaneously and paradoxically, it calls to question what exactly is the purpose or the meaning of a compulsive act, and why it had to be performed in this exact manner.

This functionalist framing of compulsive interactions requires reconceiving one's action according to a much more detailed register than what other-than-compulsive bodily action is generally alluded to. Rather than practices like 'doing groceries', 'reading a book', 'making a trip to the city centre', attending to compulsive interactions rescales how purpose becomes configured to bodily performance. This became particularly apparent from Dylan's mobile eye-tracking session, during which he examined all his movements as and when they happened in detail. He had decided on tidying his room and weighed all his actions against this overarching goal. When he picked up a t-shirt, he then looked at it, asking "why do I pick it up", he realised he could not decide whether this was helpful towards tidying his room, or if this was compulsive, as it could be considered to have worked. Thinking through why he had picked up the t-shirt, he realised he could have done multiple other acts (e.g., kicking it aside, picking up something else first, leaving it on the floor) that would be equally valid and invalid in working towards the goal of making his room tidy. Applying a functionalist ethics to his compulsive interactions meant that Dylan could no longer be certain as to what was actually intended and what was not, nor could he organise his life around elaborate practices, such as tidying his room, as it became a long string of short acts. Such an understanding of completing a task is incredibly complicated, and the chances of tidying his room derailing were very high, as it could happen with any act or motion. Upholding such a functionalist ethics based on some compulsions and living with the confusion that is produced by it places a great demand on him:

> Yeah, concentration is a bit of a thing really, it's because of which I have difficulties tidying my room and all (...) At a certain point you don't have the ability to concentrate, and you're just ... idling[5] in your room (...) that means that you're just taking hours doing nothing.

Therefore, any purpose or functionally beneficial outcome that some compulsive interactions seem to have can never be inherent to such act and will always have been applied or realised retroactively. Indeed, if compulsions could be deemed purposive and functional without upfront knowledge of their purpose and functionality, the whole premise of functional, purposive action can no longer be valid as purpose and function is defined before commencing with the act. This makes understanding one's bodily action in accordance within these larger 'practice-chunks' not only unhelpful, but also potentially damaging. As practices, such as baking a cake, cutting the grass, and putting away laundry are important compartmentalisation elements in a given day, also for recounting accomplishments during the day, taking hours to tidy one room encourages destructive self-reflections.

Compulsivity confusing the experience and sense-making efforts of other-than-compulsive action calls for a re-examination of the usefulness of practices as human geographical exploration of everyday life as well. In analytical terms, compulsive interactions unveil a certain degree of artificiality of practices. Whilst they capture the performance and embodiment of everyday life in some sense, these practice-chunks do obfuscate how smaller acts negate them and depict a much messier version of everyday life. For instance, Bill chooses a parking spot when arriving back home after we have been to the supermarket during his observation. He stops the car when we circle into the mall to point out how the situation compels him to act compulsively when choosing a parking space. Thinking with practices, parking the car becomes part of a 'commuting' practice (after Schatzki 2010), which obscures the immediacy and affective resonances of the parking act itself. Whilst such an act can be attributed to being closer to his house, it can also reflect feeling compelled to close the last gap in a row, as Bill contended. He has to drive his car into that spot to keep a rhythm of free and occupied spots or red and grey cars going, or to make the particular turn circle associated with this particular spot. In fact, compulsive acts interspersing practices demonstrate how practices are the smallest category of human action that can be attributed meaning, which retains a humanism to an otherwise potentially mechanic analysis of human action. However, such analysis excludes the consideration of human action that falls outside it, wrongly suggesting that people can know everything they do upfront, even when recognising that bodily action is emergent and constituted by virtue of the bodily situation (Schatzki 2010).

Whilst compulsive interactions cannot be inherently functional or serve a purpose, Dylan's confusion in attempting to signal compulsivity when picking up his t-shirt does call into question how other-than-compulsive bodily action serves the purpose of the practise or counts as meaningful. This does not mean that purpose and meaning are less viable structures to help guide everyday life, and that they hold less analytical truth. Nor does it mean that the bodily actions sustaining practices are always purposive. What it does mean is that compulsivity can underpin bodily action to a much higher extent, and simultaneously be a continuous part of everyday life and go entirely unnoticed, also because there is no *accepted* frame of reference to understand it through (see Chapter 9). Rather than centring the realities of embodied everyday life as practice theories suggest, it obscures embodiment that does not sustain meaningful aspects of life. It also perpetuates the idea that human beings act rationally outside their emotional life. Indeed, taking compulsivity seriously demands a further reconsideration of the usefulness of the reductive accounts of behavioural understandings of embodied life and oversimplification of what people pursue at any one moment and to what extent they have predetermined goals when embarking on an engagement. In turn, this upholds the abject status of compulsivity and affirms its deviancy, which can further alienate those who have been medicalised for their compulsive acts.

In human geography and as part of the broader social sciences, the use of practices has been an important methodology to consider how meaning does not only derive from the places in which extraordinary, spectacular, and shocking, as well as traditional and spiritual moments in life unfold, but also in the places in which the 'mundane' everyday plays out (Thrift 2004, Bondi et al. 2005, 2008, Anderson and Harrison 2010, Macpherson 2010, Degnen 2013). Despite feminist geographers and proponents of non-representational theory calling for the consideration of bodily performance more broadly, attention to bodily movements and motions has mostly been drawn to those that are meaningful, leaving out those that fail to adhere to structures of meaning. Compulsive interactions as bodily action that is seemingly impossible to derive meaning from then stretches the limits of non-representational dimension of life and more fully 'occupies' affect theories. In this manner it allows to understand more fully how body-world formations create the incessant forward movements that make up life and create the pre-conditions for more meaningful dimensions of life (see Beljaars 2020).

Explanatory gaps

With a handful of compulsive acts that participants did consider in accordance with structures of meaning or functionalism, *all* other compulsive acts for *all* other participants escaped sense-making processes entirely.

For these acts there was no explanation, neither did one emerge during performance of the act, nor in retrospect, they simply remained outside all kinds of comprehension. Whilst compulsivity has been, and for some continues to be, an aspect of their lives that is difficult to make peace with, attempts to make sense of compulsive interactions had ceased for the most part. The absence of a reason for compulsions as and when they happened was not considered a strict anomaly anymore. Compulsive interactions taking place was certainly not experienced as a shocking event in narrations of recent times, nor did they appear as such during the research meetings.

The ease with which the non-comprehensible compulsions threaded through the meetings can be characterised by the following. During Elisa's mobile eye-tracking session, she worked in her back garden, as this activity tends to incite many compulsive tendencies and acts. As I sat on a garden chair at the table, I asked if the order to which she worked her way along the plants and scrubs mattered. She got up and looked at the remaining sections of the garden, pointing at the vegetation. She elaborates:

Elisa: I will first finish the bushes before we continue with eh ... hoeing

DB: Because that isn't

Elisa: No, no, no, no, that's a different system *laughs* I start with that upper corner, that lower corner, that upper corner, and then I'll do this side and then I'll continue there, and then I'll go here. Yeah sorry *laughs*, no explanation! *laughs*

Such an explanatory gap arguably implies that psychological involvement is limited if not absent, and explanations for the compulsive performances need to be found via different routes. Precisely the failings of the psycho-analytical narrative could support this, as the gap posed a possibility for psychoanalysts to fill it with theories of sexual repression, as the acts could relatively easily be read in such terms. There was very little limitation to the sort of meaning that could be attributed to compulsions (and tics), despite often seeming very far removed from the movements, objects, and social situations that constituted the compulsion or tic. For instance, Ferenczi associated palipraxia (imitation of gestures of another) with irony, whereas current psychological theories tend to connect this to being impressed with the other person; not to mention his interpretation of compulsive acts as infantilism and narcissism[6].

As interpretation of compulsions is possible and limitless as there are no assistive explanatory systems, hence any interpretation seems viable. For instance, the compulsion to align a pluck of hair that sticks up from another person's head with a vertical line in the wallpaper behind them during a conversation: when the other person moves and disrupts the alignment, the first person needs to reposition to reinstate the alignment. Interpretive analysis could focus, for instance, on the conversation style and deem the act palipraxia, or on the social relation between the two people, or on an obsession with hair. Therefore, *getting it wrong* through an interpretive framework is the overwhelmingly likely possibility. Employing interpretive theories and assigning truth to these acts through representational assertions thus poses a real danger to those who are suffering, not only in having to understand themselves in these – often embarrassing – terms, in particular those offered by Freudian and Ferenczian psychoanalysis, but also in terms of treatment options on offer and in development. Further effects of such interpretive analysis can involve invoking and upholding the normative structures that feed and maintain stigma.

The void that remains and seems to be illusive to capture in existing styles of analysis does not mean that nothing remains to be done. Rather, it should be questioned why certain explanatory structures are insisted on, even after continued unsatisfactory delivery of explanations. What cannot be located and what escapes our view and focus? Viewing compulsivity afresh does not mean that experience as underpinning analytical frameworks of compulsive acts should be shunned altogether. Indeed, rather than a representation or approximation of the cognitive – and to a certain extent – affective structures that govern experience through its articulation in words, notwithstanding the fact that words and feelings mutually inform each other, considering experience and the severely limited options as unquestioned truth and real worlding of compulsive interactions can produce valuable new understandings of them. The experience of those who live through the sensations and movements that constitute compulsive acts allows to immediately and unreservedly inform and guide such analysis, in ways psychoanalytical – nor other – approaches have not been able – or have not been deemed necessary – to do. Such analysis would continue to allow distinguishing between types of compulsions, different circumstances and situations, as well as between individual interactions.

Examining compulsive interactions thus highlights several confusions that emerge from the unsatisfactory explanations that familiar structures of meaning could offer. Consequently, this framework is insufficient in explaining how compulsive interactions and compulsivity are constituted more broadly. This, in combination with compulsivity and compulsive interactions produced significant suffering in the overlapping physical, mental, and socio-normative realms of life for the person involved, as well as the deviation from social norms and values about bodily action, inciting consideration of the phenomenon as problematic. Medicalisation of the acts has produced an alternative conceptual framework that marks the social world as irrelevant, and moves away from explanation based on interpretation and relating it to the person or their history. Deeming all compulsions and compulsivity intrinsically meaningless, the neuropsychiatric sciences formulate a biological argument that seeks the constitution of compulsivity and compulsive interactions in the brain.

Notes

1 Please note that ellipsis without parentheses denote pauses and ellipsis with parentheses denote omissions throughout the book
2 With thanks to Jo Bervoets for the suggestion to emphasise this element.
3 Hysteria was considered an affliction that affected mostly women for a broad range of mostly social reasons with physical manifestations in the brain. Fiercely debated, Hysteria was understood to be caused by a lack of sexual gratification (Charcot in Veith 1965; Freud 1896), childhood sexual trauma (Freud 1896), "forbidden wishes and longings" (Swetlow 1953) (see Maines 1999). It is no longer recognised as a diagnosis of mental illness.
4 Translated into English in 1907: Meige, H. and Feindel, E. (1907) *Tics and their Treatment*, with a preface by Professor Brissaud. (Trans & Ed: S.A.K. Wilson). New York: William Wood and Co.
5 Original Dutch expression: *'lanterfanten'*.
6 Another example of an interpretation of compulsions as signs of sexual repression can be found in a 1916/1917 lecture by Freud in which discusses one of his cases. He interprets all the small gestures and acts of a 19-year-old woman during her compulsive bedtime ritual as sexual in one way or another.
   (Text source: Freud, "The Sense of Symptoms", in James Strachey, ed., The Standard Edition of the Complete Psychological Works of Sigmund Freud. Vol. 16, London: Hogarth Press, pp. 264–269.) The excerpt can be read online: https://www.ocdhistory.net/20thcentury/freud.html

References

Anderson, B. and Harrison, P. eds. 2010. *Taking-Place: Non-Representational Theories and Geography*. Farnham: Ashgate.

Beljaars, D. 2020. Towards compulsive geographies. *Transactions of the Institute of British Geographers* 45, pp. 284–298.

Bondi, L., Davidson, J. and Smith, M. 2005. Introduction: Geography's 'emotional turn'. In: Davidson, J., Bondi, L. and Smith, M. eds. *Emotional Geographies*. Ashgate, Aldershot, pp. 1–16.

Degnen, C. 2013. 'Knowing', absence, and presence: The spatial and temporal depth of relations. *Environment and Planning D: Society and Space* 31, pp. 554–570.

Deleuze, G. and Guattari, F. 2004 [1980]. *A Thousand Plateaus. Capitalism and Schizophrenia*. London: Continuum.

Fenichel, O. 1945. *The Psychoanalytic Theory of Neurosis*. New York, NY: Norton and Company.

Ferenczi, S. 1921. Psycho-analytical observations on tic. *International Journal of Psychoanalysis* 2, pp. 1–30.

Freud, S. 1896. "The aetiology of hysteria", *The Standard Edition of the Complete Psychological Works of Sigmund Freud*. Vol. 3, trans. Strachey J., London: Hogarth Press, pp. 189–208.

Freud, S. 1953–1974. "The sense of symptoms". In: Strachey, J. ed. *The Standard Edition of the Complete Psychological Works of Sigmund Freud*. Vol. 16, London: Hogarth Press, pp. 264–269.

Hollenbeck, P.J. 2003. A jangling journey: Life with Tourette syndrome. *Cerebrum* 53, pp. 47–61.

Kushner, H.I. 1999. *A Cursing Brain? The Histories of Tourettes Syndrome*. Cambridge, MA: Harvard University Press.

Lerch, O. 1901. Convulsive tics. *American Medicine* pp. 694–695.

Macpherson, H. 2010. Non-representational approaches to body–landscape relations. *Geography Compass* 4, pp. 1–13.

Mahler, M.S. and Gross I.L. 1945. Psychotherapeutic study of a typical case with tic syndrome. *The Nervous Child* 4, pp. 359–373.

Mahler, M.S. and Rangell, L. 1943. A psychosomatic study of maladie des tics (Gilles de la Tourette's syndrome). *The Psychiatric Quarterly* 17, pp. 579–603.

Maines, R.P. 1999. *The Technology of Orgasm: 'Hysteria', the Vibrator, and Women's Sexual Satisfaction*. Baltimore, ML: The Johns Hopkins University Press.

Meige, H. and Feindel, E. 1907 [1902]. *Tics and Their Treatment*, trans & ed: Wilson, S.A.K., New York, NY: William Wood and Co.

Schatzki, T.R. 2010. *The Timespace of Human Activity: On Performance, Society, and History as Indeterminate Teleological Events*. Lanham, MA: Lexington Books.

Shapiro, A.K. and Shapiro, E. 1968. Treatment of Gilles de la Tourette's syndrome with Haloperidol. *British Journal of Psychiatry* 114, pp. 345–350.

Simondon, G. 2020 [1964]. *Individuation in Light of Notions of Form and Information*, trans. Adkins, T., Minneapolis MN: University of Minnesota Press.

Swetlow, G. 1953. "Hyterics as litigants", in *Bulletin of the Medical Society of the County of Kings*. c.f. Maines, R.P. 1999. *The Technology of Orgasm: 'Hysteria', the Vibrator, and Women's Sexual Satisfaction*. Baltimore & London: The Johns Hopkins University Press.

Thrift, N. 2004. Intensities of feeling: Towards a spatial politics of affect. *Geografiska Annaler B* 861, pp. 57–78.

Thrift, N. 2008. *Non-Representational Theory: Space, Politics, Affect*. London: Routledge.

Veith, I. 1965. *Hysteria: The History of a Disease*. Chicago, IL: University of Chicago Press.

Wilson, J.C. 1897. A case of tic convulsif. *Archives of Pediatrics* 14, pp. 881–887.

2 Complications

Neuropsychiatric rationalisations

As the experiential rationalities that govern other-than-compulsive bodily action are insufficient to understand compulsivity and compulsive interactions, the current-day neuropsychiatric and current-day non-psychoanalytical clinical sciences present an alternative framework for understanding this aspect of people's life. In effect, this medicalisation of compulsion is another effort to rationalise the phenomenon. Rather than centring experiential systems of logic, this effort is built on biological and neurocognitive arguments. By no means can the neuropsychiatric, biomedical, and clinical sciences be seen as putting forward one undisputed understanding of compulsions, and the these sciences of compulsivity form a complex landscape filled with differences and internal disagreements, as will become clear in this chapter. However, as these sciences are underpinned by positivism, they share principles of what counts as valuable knowledge and how valuable knowledge should be created, and for that reason they are discussed in combination. Whilst these principles have led these sciences to produce extensive insight into compulsivity, as this chapter points out, they have also led to neglect certain dimensions, a dismissal of particular sources of knowledge, and a discouragement of other kinds of knowledge creation. As a result, confusions surrounding compulsive interactions partially resolve and partially shift into new explanatory territories.

In order for the neuropsychiatric, biomedical, and clinical rationalisations to formulate explanations as to why compulsivity can be part of human life and why compulsive interactions take place, the phenomenon has to go through a number of transformations. These transformations present another set of origins and set of problem definitions, many of which are not immediately, or at all, derived from the experiences of the people who perform compulsions. Rather, they become derivatives of diagnostic and neurological categorisations and different analytical epistemes. The onus of this rationalisation then does not fall on *why certain compulsive acts* take place, but on *how compulsivity is an abnormality* and how *this body* is *affected*. Indeed, this reframing reflects the research philosophy underpinning the primarily deductive approaches that produce 'objective', 'outsider' knowledge of compulsive acts in quantitative terms of difference in *degree*,

DOI: 10.4324/9781003109921-3

but not *kind*, nor do they necessarily engage with the *socio-spatial circumstances* under which compulsions are performed.

This chapter traces the neuropsychiatric rationalisation of compulsive interactions and considers how it affects the confusions that people with Tourette syndrome who perform compulsive acts experience. It does so by examining four interconnected transformations and the onto-epistemological philosophies that underpin them as well as their implications for the ways in which the phenomenon has been rendered and treatment has been mapped onto it. This brings to light how certain confusions surrounding these compulsions have, at best, been lifted or shifted, have been neglected, or, at worst, have been exacerbated.

Transformation 1: Pathologisation

The neuropsychiatric scientific rationalisation of compulsive realities involves clinical diagnosis and starts at the point that compulsions are so problematic that professional help is required to cope with them. For many people, this rendition becomes an important – if not the most important – framework of understanding when they are diagnosed with Tourette syndrome, or as Joe puts it "I had it my entire life of course, but officially I know it for only four years". In lieu of failing to find sufficient answers to their compulsive experiences and the desperation that accompanies it, this new formal rendition offers a persuasive new understanding that could dissolve many confusions, or at least bring (partial) answers. Its comprehensive conceptualisation offers an explanation of many aspects of people's lives that extend beyond compulsive acts. The process of diagnostisation and eventually receiving the diagnosis of Tourette syndrome radically changes how they consider compulsivity, and how compulsive acts are perceived and even how these are experienced.

The pathologisation of a particular aspect of their lives introduces a normative division between healthy or normal engagements and unhealthy or abnormal ones. All compulsive interactions are thus grouped by virtue of it not adhering to the behavioural rationalities that govern what is defined as 'healthy', 'intended', and 'rational' behaviours. Indeed, for many research participants the Tourette's diagnosis had become a shorthand for alluding to their compulsive bodily movements and engagements. After his first mobile eye-tracking session, Dylan and I watch the recordings, and he considers his compulsions: "it all doesn't look very Touretty, I think", but he cannot quite put his finger on what it is that makes movements themselves particularly 'Touretty'. In line with Dylan and many others when examining their compulsivity, Tomos invokes his compulsive sensibilities as entirely self-contained psychological process when he explains his daily struggle: "then Tourette's really takes power over me I'm afraid". Whilst the clinical and experiential vocabularies differ in constitution and purpose, as well as remit and scope, in capturing compulsivity, the participants often merged these vocabularies in the expression of their experiences. Hence, with the diagnosis, a new priority is placed on the shared seemingly unintended and irrational dimension of compulsions, which reduces the need for people to consider all compulsive acts as individual acts in the adoption of the clinical rationalisation of compulsive acts. The new diagnostic framing of these acts then helps to reduce their own confusions about the phenomenon and provides them with an appealing 'way out'; something that sits in between themselves and the acts.

The transformation of lived experience and knowledge of one's body to the clinical rendering of one's compulsive movements can be clarifying. Rhys demonstrates how it takes away confusions. For instance, Rhys' incessant sniff he had from a young age, and what he believed had been the sign of him having a year-round cold, was actually deemed to be a compulsion; a 'vocal tic' in the medical terminology. This refiguration of his body lifted worries over his immune system and helped him better understand his bodily movements. However, understanding the pathologisation of one's compulsions does not always go smoothly. The pathologisation of Lowri's acts left her struggling to understand which ones were problematised. She had expected to better understand her unwanted touching and ordering tendencies through acquainting herself with the clinical vocabulary after her diagnosis. However, she found it to be very confusing because of its narrow and rigid definitions and "was unsure if [she] was relieved at that point". It led her to closely examine all her movements, habits, routines, rituals, and preferences that she could not quite explain, and consequently feeling ashamed for "not knowing" that a particular movement she used to do "was a tic". In effect, the clinical rendition of her movements that she needed to adhere to led her to question her knowledge of her body as gathered throughout her life.

The pathologisation and diagnostisation of compulsive interactions change people's understanding of compulsive interactions because both processes change how they relate to these acts, their body and themselves. Their compulsive interactions – or tics to be more precise – were pathologised, but like in other illnesses, *they as individuals* were diagnosed. In the interview Joe remarks that he struggled with this transformation, but Sage in her interview welcomed it:

> It was rather difficult to take that first step, to ask for help, you know. You get a stamp 'he's faulty, so he's probably not right in the head'
>
> (Joe)

> I recall feeling incredibly relieved, because I knew already that I didn't do things on purpose (…) like an 'I told you so!' confirmation: 'I can't help it', so that was all good really.
>
> (Sage)

Joe's fear reflects a fundamental issue that is well developed in health sociology, disability studies, and the critical medical humanities; pathologisation of a person and diagnostisation of a body risks the reduction of people as victims of their own biology; it renders them more predictable and takes away agency that is granted to undiagnosed others. Also, it opens the door to a stigma-induced overclaim of people's incapacities, for instance in terms of self-organisation. Subsequently, it lowers the standards to which undiagnosed others are held, how they can be treated, and what merits intervention.

Tying her diagnosis to her social relations, Sage reflects on ongoing disputes with her mother about the nature of her compulsions, as her mother would not believe Sage could not help but performing them. The diagnosis then changes individuals having to take responsibility for their compulsive bodily movements to the admission that there was a problem in their body, thereby shifting the location of the problem away from themselves, and onto their diagnosed body, and/or Tourette's (Schroeder 2005, Sandle 2012, Bervoets forthcoming, Bervoets and Beljaars, in review). This transforms compulsivity as confusing acts – 'why do I perform this act?' – to a problem of the individual – 'something is wrong with *me*' – and to a biological problem of their brain – 'my brain is faulty'. Paradoxically perhaps, the focus of the diagnosis on the individual and its biologisation of the compulsive interactions that it incites does help to lift confusions rooted in questions about the self and one's desires.

Transformation 2: Biologisation and neuroscientific logic

Another transformation that builds on the pathological one entails the biologisation of the compulsive interactions. The compulsive tendencies and the interactions that had puzzled Joe and others continue to shift from a personal crisis of intention and meaning to a normative behavioural enigma and then to a biological defect. In particular, the neuropsychiatric sciences formulate compulsivity as arising from a malfunctioning brain and the distorted processes that make up the nervous system. Following Robbins et al. (2012: 81), the combination of the broad set of involved neuropsychiatric perspectives entail a biological approach that is based on 'neurocognitive endophenotypes', "whereby changes in behavioural or cognitive processes are associated with discrete deficits in defined neural systems". In other words, this reflects the notion that what bodies do is a direct expression of the functionality of the brain, so neuroscientific logic holds that if people do things that are considered abnormal, there must be something wrong with their brain. This logic reflects a conceptualisation of the problem fundamentally as a deficit.

Tics and compulsions were considered to having a distinct neurological cause already in the late 1800s. As part of the eugenicist ideology that drove many sciences (Dowbiggin 1991), Georges Gilles De la Tourette (1885) and Jean-Martin Charcot (1887–1888, c.f. Kushner 1999), saw them as manifesting problems with the nervous system that signified a degenerative inheritance that was caused by alcoholism and immoral behaviour by past generations (Kushner 1999). Currently, compulsive interactions are considered to be "caused by a defective metabolism of the neurotransmitters in the brain" (2020 ICD-10-CM Diagnosis Code F95.2): particularly structurally and functionally involving the basal ganglia, thalamus, prefrontal cortex and the cerebellum (for an overview see Ramkiran et al. 2019). This clear premise for the organic rooting of compulsions has been arrived at through different methods of disrupting their occurrence.

Understanding compulsive interactions as emerging from a problem in the brain thus permits the neurosciences a variety of interventions, notably physically, electronically, and chemically. Any kind of reduction in tics and compulsive interactions is deemed as a successful interference. Chemical interference has developed from the anti-psychotic drug Haloperidol. Kushner (1999) explains that this is a tranquilliser originally utilised to handle 'unmanageable' psychiatric patients that can induce Parkinson-like movements if taken in doses that are too high. The reduction in compulsions were considered to break the neurobiological chain in the production of the urge to perform them, but not cure the brain, as stopping taking the medication would increase the tics again. This is the same for current-day medication that includes neuroleptic and a-typical antipsychotics (e.g. aripiprazole, risperidone, olanzapine, and ziprasidone), dopamine receptor blockers or first-line antipsychotics (e.g. haloperidol, pimozide, and fluphenazine). Whereas atypical antipsychotics are preferred, what medicine is prescribed depends on the kind and multitude of diagnoses, as well as the severity as established with clinical tests, such as the Yale Global Tic Severity Scale (YGTSS) (Leckman et al. 1989).

No specific medicine has been developed for urge-driven conditions, therefore these drugs have pervasive effects (see Lombroso and Scahill 2008, Shprecher and Kurlan 2009), which also tend to cause an increase of many unwanted and other harmful phenomena that have been rendered as 'side effects'. They included muscle spasms, zombiism, restlessness, lethargy, phobias, suicidal thoughts, and can be so powerful that many have to stop treatment. Arguably, this treatment option is like the method of a shotgun: when shot and one shell hits the target, it could be claimed as a success, independent of the other elements shot by the other shells. Although the workings of antidepressants are related to different serotonin systems in the brain, it is not entirely clear why certain this type and other types of medication have certain effect on one person's tics and compulsions, whilst another person can have very different experiences.

In the 1970s neurosurgery entailed lobotomy in which lesions were caused in the thalamus to permanently disrupt brain structures and was found to reduce tics, and probably compulsive interactions, for most people (Kushner 1999). Since the 2000s, the invasive, but ostensibly reversible, Deep Brain Stimulation (DBS) seems to reduce the 'superfluous' movements for at least up to three years (Kimura et al. 2021), probably also including urge-driven compulsive interactions through electrodes implanted in the brain offering variable electric *stimulation* (see Xu et al. 2020). Despite the research that seeks to map the production of tics in great detail, Kimura et al. (2021) argue that the electrodes can be placed in a wide variety of brain parts and yield the same results. The new physically non-invasive brain oscillation interference technique reduces tics by stimulates the cortical motor areas in the brain with an altered pattern of electric stimuli that is introduced to the Median Nerve via an electrified wristband (Morera Maiquez et al. 2020). It is important to keep in mind that any treatment that targets the brain is deemed successful when it reduces tics, including compulsive interactions and the plethora of severe side effects[1]. Lobotomy was found to reduce tics for less than 2 years but returned more severely afterward (Asam and Karrass 1981). For DBS, long-term effects are still largely unknown (but see Smeets et al. 2018) although it does tend to reduce tics for the overwhelming majority of those treated in the short term. For the new brain wave interference technique both long-term effects on compulsions and other effects are also as yet unknown (Morera Maiquez et al. 2020)

The effects of antipsychotics and neurosurgery on the manifestation of compulsions does not necessarily prove that they emerge from damage to the brain, rather that the chemicals, electrodes, and wristband interfere with their appearance through their alteration of the nerve stimulations or dopaminic regimes in the nervous system. Neither does it mean that the workings on tics necessarily indicate that there is something wrong with the brain. However, framing compulsivity as a neurobiological 'defect' in 'inhibitory' structures does not only open up the possibility to fix it, it also favours such fixture to be of the neurobiological kind. This reinforces and expands the importance of the roles of neuroscientific knowledge in understanding the phenomenon. In effect, it implies that a reduction in compulsive interactions equals a better functioning brain, and that an interference achieving the absence of compulsions for at least a year (in accordance to the diagnosis) signifies an optimal outcome for Tourette's as a disordered condition.

As with any science organised around positivist knowledge construction and mobilising platonic essentialism in its conceptualisation of phenomena, the life sciences of Tourette syndrome understand compulsive interactions as symptoms or signifiers of an underlying problem that causes them to manifest. As explained elsewhere, compulsions as conceived as biological entities grants compulsivity and the Tourette's diagnoses an ongoing ontological presence, rather than an intermittent one if the compulsive acts had been given ontic essence (Beljaars 2020). This onto-epistemology that guides these sciences thus has a strong transformative power on the conceptualisation of compulsive interactions. It follows that such understandings are built on dimensions of the phenomena that are, or otherwise become, measurable and quantifiable, brought into relations that are correlative – and not causal – in nature. The studies, as well as the DSM, prioritise observational methodology, assuming a universality of the human body, and reinforce findings through 'evidence-based' regimes of knowledge creation that only

\

recognise similar kinds of data as valuable input for new research. These sciences therefore exclude apprehending compulsive interactions in all their complexity (Clegg et al. 2013) and take greater experiential dimensions into account (Bankey 2004, Greenhough 2011).

In lieu of Tourette's having been declared a neurodevelopmental disorder, compulsive interactions have been included on a scale of symptom 'complexity' of lesser to greater extent. In this transformation, compulsive interactions are recast as *movements*, centring the involvement of the bones, muscles, and tendons[2]. Involving multiple muscle groups, compulsive interactions are then considered the most complex and are placed at the extreme end of the scale. People diagnosed with Tourette syndrome are considered to perform 'simple tics' in childhood and start to perform more complex compulsions when they age into adolescence (Bloch et al. 2006, Bloch & Leckman 2009, Groth et al. 2017). However, simple tics remain the hallmark symptom of the diagnosis from which it follows that the Tourette's condition is considered to go into remission (see Bloch et al. 2006, Rothenberger and Roessner 2019), and that it might well be possible that compulsive interactions are too complex to be recognised as symptoms. As they may take longer than the very brief instance of a tic, look like, or be made to look like 'normal' movements, or occur outside medical and clinical diagnostic, treatment, and research spaces, no research effort into these relatively extremely complex acts would be mandated according to the model's theoretical logic, despite its capacity for such research. In practice, the life scientific epistemological tendency to create and utilise stable constructs of phenomena, such as through disinhibition, cannot account for the acknowledged inherent slipperiness of the symptom expression that is used to allude to compulsive interactions.

The transformation that the neuropsychiatric, biomedical, and clinical rationalisation of compulsive interactions induces through their biologisation reconceptualises them as complex in biomechanical, rather than psychological, terms, and it obscures them by conceiving them as merely an effect of a brain problem. Whilst this strongly reductive account of compulsive acts does not offer intricate answers about compulsive interactions, it can offer a helpful way to make sense of inexplicable elements of their occurrence. Dylan's sense-making processes of having to do acts he does not intend to do are strongly based on his knowledge about their biological dimension. Exploring how compulsions feel he brings our discussion to the workings of neurotransmitters and how they are a reality for him:

> Dylan: You kind of feel that there's something not right in your nervous system. The only reason it's not right is of course because there are too many neurotransmitters, and not enough here and there … Even that's not even known, ehm … the argument at the moment is that you have too much dopamine, yeah okay … then that's how I shall put it; I have too much dopamine in my ass[3] … but if it actually works like that, if that's

<u>really</u> true ... and seratonin, noradrenalin are coming back, so they're included. But those are the most well-known ones ... ehm ... it goes a lot deeper and, actually, if that's what you have, you could medicate it.

DB: yeah, but if you then exactly know if it ... is secreting too much neurotransmitters ...

Dylan: You feel something is not right, ehm ... if you transmit a signal over a particular nerve, then you can make it right in a way, so it's often a movement stimulus ehm ... A movement stimulus thus means that you flex a muscle until the sensation has passed through the nerve, which makes it more right again.

The biologisation of compulsion and compulsive processes that precede the acts is then positioned as whole explanation of the phenomenon or as partial explanation that fit the gaps that are left by the incomplete sense-making exercises. In other words, the focus on inhibition puts forward the idea that acts can be good or bad, and that therefore the problem just is a lack of neurotransmitter-mechanics to impair the 'bad'[4]. It is a poignant reality that takes away the need to ask questions about *this* act, because it effectively shifts the focus to *how* any compulsion happens in the first place.

Transformation 3: Erasure of performative difference

Another transformation of considering compulsive interactions that stems from the rationalisation of the neuropsychiatric, biomedical, and clinical sciences involves a change in the possibility for certain questions that can be asked. As a result of the biologisation of compulsivity analytically situating the brain and further nervous system as the causal focal point, bodily movements are understood as little more than effects of the brain problem. Their rendition as 'symptom' is more important than the intricacies of the compulsions themselves. What these movements look like does not necessarily matter in the neuropsychiatric scientific analysis: the difference between Siôn having to compulsively reorder the dishwasher and Cai having to compulsively pick up his parents' cats has no analytical meaning or value in this rationalisation. Therefore, these understandings offer very little insight into the kinds of compulsive acts that people do. This deindividualisation transformation of compulsive interactions emerges from the biologisation in accordance with its positivist onto-epistemological principles and diagnostic procedures and their related categorisation exercises.

As positivist epistemologies do not require a sensitivity towards difference in kind and as there seems to be no limit to the variation between tics and compulsions, medical and clinical literature understand the instability of the symptom group as a given. With the exception of a short period in the 1930s, neurobiological sciences have not endeavoured to find patterns of the recurrence of particular compulsions (Kushner 2008). Therefore, tics and compulsions are conceived of as 'highly idiosyncratic' (e.g. O'Connor et al. 1994, Verdellen 2007) which remains unchallenged. On this basis, research participants in positivist life scientific research – with the exception of those partaking in phenotype studies – are often requested to quantify their experience of having to do a compulsion as a singular experiential entity. According to the participants to this study, this quantification is one of the most profound transformations their compulsive interactions go through to fit the life scientific episteme.

Mina's professional medical knowledge and commensurate familiarity with clinical vocabularies and diagnostics had led her to consider how compulsivity should be studied based on her own experiences. She had been keen to talk through the clinical construction of her experiences during all our meetings, and during her second eye-tracking session, we discuss the implications of the genetic overlap between 'compulsive disorders' and 'schizophrenia' which she had read an article about. She had proposed to make a drawing as it would bring out her compulsive tendencies in great detail. Getting increasingly frustrated with how the charcoal lion emerged on the paper in front of her, she speaks slowly and deliberately relating her compulsions to the concepts of psychosis and addiction:

> Why it didn't really surprise me, but that would – it's quite unscientific what I'm saying – because it's so persistent and because these psychotic disorders seem so organic, I thought it really wouldn't surprise me; it's rather psychotic, I think, those compulsions. (...) It's something that drives you. I mean, it's outside your control. If you have a psychosis, and without having any say over it's like when you have compulsions ... And that it's also something that you almost can't have a say over. Then addiction is also something that just gets worse all the time ... So it's something – I would say – that it's outside yourself and controls you, and that's also in psychiatry. Maybe that's why I keep thinking about it. I think it's really just a kind of psychosis, maybe they should see it more like that.

In suggesting how the fluidity of the unfolding and the oppressive feeling of psychosis and addiction and compulsivity are very similar in her experience, she paints an intricate picture of how compulsivity has an ongoing complex presence. She depicts a phenomenon that is difficult to even capture in words, let alone in the rigid and categorical neuropsychiatric description that requires it to have a traceable, universal, and consistent numerical existence. The fact that so much is lost in this transformation renders the neuropsychiatric conception unhelpful, which makes her doubt the usefulness of the treatment options on offer.

In addition to the necessity for rendering compulsive acts quantifiable, the neuropsychiatric conceptualisation requires diagnosed people to unify all their sensations and actions and discuss them on the same terms. Indeed, the life scientific 'unit of calculation' is a diagnosed person or the presence of symptom, not an act itself, which makes the compelled movements entirely a function of the individual. Therefore, where not further specified, blanket statements are made about different movements and acts (in addition to different sensations) under the heading of 'symptoms' because it is the same person who performs them (e.g. van der Salm et al. 2012). In turn, this confirms and emphasises the requirement for the person to be diagnosed with Tourette syndrome.

Diagnosis is premised on a deductive method of determining what disease a patient has; criteria forming one diagnosis are therefore always different from another diagnoses. Individual symptoms can occur with other diagnoses; pain for instance, but the combination of the symptom collection is unique to a diagnosis. The diagnostic criteria for Tourette syndrome consist of two motor tics, for example eye-blinking, shoulder shrugging, or nose scrunching, and one vocal tic, such as sniffing, coughing, or uttering a sound or word, for at least a year with an onset prior to the age of 18, and not as a result of medicine or other drugs (DSM5). The complex and intricate issues that have led people like Mina to seek diagnosis, including compulsive interactions, nonetheless, the diagnosis shifts motor and vocal tics into the central focus. As these bodily movements are most strongly problematised and foregrounded in treatment options, it incites a diagnosed person to prioritise the experience of these particular bodily movements over others (see also Kushner 1999). In effect, the clinical rationalisation of bodily movement introduces a hierarchy of more and less important acts that background compulsive interactions.

Compulsive interaction is not an official symptom category of Tourette syndrome, as the movements are regarded as acts, which reflects the ontic focus on the universal biological body. The clinical capture of compulsive interactions emerges from 'phenotype' studies that render visible and examine the relations between bodily movements and brain divergence by (re)producing movement categories (e.g. Cath et al. 2001, Worbe et al. 2010, Ferrao et al. 2013). These studies address the demand for practical distinction possibilities between the symptomatologies of Tourette syndrome, Obsessive Compulsive Disorder (OCD), and Attention Deficit Hyperactivity Disorder (ADHD) diagnoses to improve the capture of these diagnoses, reduce misdiagnosis, and signal 'comorbidities' – the clinical recognition of people having multiple diagnoses. Unclear how these categories are produced exactly, they are reified in deductive (clinical) tests, such as the most widely employed Yale Global Tic Severity Scale (YGTSS) (Leckman et al. 1989).

The clinical categories that capture compulsive interactions involve 'touching' (e.g. pressing one's finger into the corner of a table, or clasping a mug), and include 'tapping', 'rubbing', 'clapping', and 'picking' (e.g. removing a flower from its bed, or a leaf from its stalk), 'ordering' (e.g. grouping similar objects, or repositioning objects into a new composition) that includes positioning, arranging, 'symmetry behaviour', and 'evening-up performances' (Cath et al. 1992, 2001, Rosario-Campos et al. 2001, Alsobrook and Pauls 2002,

Mansueto and Keuler 2005, Palumbo and Kurlan 2007, Robertson and Cavanna 2007, Worbe et al. 2010, Ferrao et al. 2013, Neal and Cavanna 2013, Huisman-van Dijk et al. 2016, Sambrani et al. 2016). The clinical capture of compulsive interactions also extends to 'paliphenomena', describing acts that involve repetition of actions (palipraxia) or sounds (palilalia): their own or those from other people, animals, and objects. Mina explains that for her this includes having to reread a paragraph in a book "because it was not read properly", and for Lowri this is typified by her having to step through a doorframe again if the first time was not done in the 'right' way.

Compulsive interactions that involve seeking or creating some kind of pattern in the broadly conceived bodily environment have been clinically denoted under the headings of 'mental play' (Cath et al. 2001, Worbe et al. 2010) or 'mental compulsions' (Williams et al. 2011). They include counting a variety of things and seeking (un)even amounts of things in a place, finger tapping on musical rhythms, as well as aligning objects and people visually (Cath et al. 2001, Alsobrook and Pauls 2002, Worbe et al. 2010). This latter kind resonates with Alan, who had to move his head to visually align a strand of my hair with a lamppost outside whilst we were having lunch after the interview. Another example of these interactions that demonstrate an astoundingly complex spatial imaginary is artist and karate instructor Shane Fistell compulsively blowing air close to Oliver Sacks' mouth because he had to 'touch' Sacks' breath with his own. This compulsion and many others figure on *The Mind Traveller* (1996)[5], a documentary following Dr Oliver Sacks engaging with some of his patients.

Compulsive interactions that have painful consequences to the person are denoted as 'self-injurious behaviours'. In contrast to the other categories, these compulsions do not pinpoint a particular kind of movement; rather they are used to indicate clinical severity of Tourette syndrome. Distinctly different from acts of self-harm, the harmful element is an *unwanted consequence*, not the purpose, of the act, according to people who have to perform them. Examples include hitting oneself in the chest, violently dropping oneself on the floor, burning or cutting parts of one's own body (Robertson and Cavanna 2007, Robertson et al. 2008). Despite these particular clinical categorisations of compulsive interactions and other acts as part of the Tourette's symptomatology, they are not treated differently[6], nor do they incite a diversification of clinical explanations as to why people compulsively touch rather than compulsively arrange, for instance.

The limited clinical capture of compulsive interactions incites a reduced acknowledgement or ill-recognition of others. In addition to the quantification of their sensations, many participants to this study affirmed that this transformation that is imposed from the outside is not always particularly welcomed. Having aligned his expressions of his compulsions in strong alignment with the biological conception, during our meetings Dylan often questioned his movements and sensations on their accuracy, and purpose. Even the way he reached for the tap during an eye-tracking session was assessed in detailed clinical categorical language. He had initially been content with the capacity of the Tourette's diagnosis to capture his experiences, but after a while I noticed how some acts and sensations could not be related to the diagnosis. As he found this absence unacceptable, he had identified a gap in the scientific literature where his experiences should be fitted:

> I actually did make up a new category that makes me think like, yeah, these are indeed motor tics. Only it doesn't consist of the movements that we know of, and I miss a category with a normal description which I named 'passive tics'. Passive motor tics … Can something like that exist? Yes of course it can exist! I mean, I am a patient, I feel that, therefore that is what it is.

We discuss what he means by that and he demonstrates what he deems a normal versus a compulsive way of sitting on the sofa:

> Passive tics at tics that in one way or another cause a pain or a pressure that you want to get rid of … That you want to get rid of by, for instance, immobilising a body part; you pull your arm very far behind your head so that you can't feel your arm anymore after a while.

The discomfort that this way of sitting invokes precludes the requirement to move his arm compulsively. Dylan expected and needed the clinical vocabulary to provide answers to, or even recognition of, his experiences to lift confusions around the phenomena, but it did not. Therefore, he needed the current set to be expanded. Nonetheless, in addition to sense-making exercises, the neurobiological and clinical rationalisations cannot explain why *this* compulsion takes place and only offer a device for description. Indeed, the erasure of differences between individual compulsive acts through the process of diagnosis is amplified by the biologisation of the acts. Hence the neuropsychiatric, biomedical, and clinical rationalisation of compulsive interactions invites people who have to perform them to discard denoting differences and accept them as idiosyncratic. This, therefore, more of less lifts the confusion, as there seem to be no answers. In a similar way, the compulsions as rationalised within the life sciences does not encourage querying if any contextual aspects could be informative of how they take place as and when they take place.

Transformation 4: Erasure of circumstances

In addition to compulsions and tics being regarded as idiosyncratic, they are deemed to be 'waxing and waning', which alludes to both the kind of compulsion performed by a diagnosed person, and the temporal variations of compulsivity in a given period. Similar to idiosyncracy of kind, the denotation of these variations in the clinical sciences have been made in the exploratory, descriptive mode serving as acknowledgement of social worlds and activities as having salience (Cohen and Leckman 1992, Conelea and Woods 2008, Woods et al. 2009, Cavanna and Nani 2013). However, as per the natural scientific pursuit of universal truths, these relations have not been subjected to a rigorous examination.

Studies do register differences in the frequency and severity of simple tics that can be observed by clinicians and close others (e.g. partners, parents) between the home and the doctors office (Goetz et al. 2001), "reading a book", "spending time with friends", and "moving to a new home" (Christenson et al. 1993, Silva et al. 1995, Miltenberg et al. 1998). Nonetheless, these studies set up activity categories that suggest the idea of certain general life situations but are too vaguely defined and lack acknowledgement of the materiality, micro-dynamics, and social fabric of situations. These are therefore too abstract to start understanding why this compulsive interaction happens now, here, and under these circumstances. Indeed, studies that do focus on 'environmental influences', such as those related to the social world, utilise questionnaires of broad categories that do not allow for nuance, nor differentiation between situations on the basis of which participants to these studies choose their answers (Leckman et al. 1993, Woods et al. 2005, Conelea and Woods 2008, Conelea et al. 2011, Wang et al. 2011, Capriotti et al. 2013).

There are exceptions in which the circumstances under which compulsions take place are acknowledged more fully. For instance, Karp and Hallett's (1996) study that is based on experiential accounts from other studies, argue that the bodily surroundings locate the starting point of the sensations that lead a person to act compulsively outside the body. These studies are almost exclusively studies that allow for more nuance through case studies (Eapen et al. 1994, Cohen et al. 2013), first person narratives (Bliss 1980, Kane 1994, Hollenbeck 2003, Turtle and Robertson 2008), and non-academic autobiographies (Wilensky 1999, Van Bloss 2006). Joseph Bliss (1980: 1347) explains this as "a mental projection of sensory impressions to other persons and to inanimate or even non-existent objects". For example, he would perceive a "firm cord running down the center line of the sheet. A need appears to apply pressure to this phantom cord by pulling" (ibid.). Bliss also describes *feeling*, not *touching* an object:

> At times there is a recurring need while writing to press the pencil point hard against the surface of the paper. A 'feel' is perceived at the end of the pencil; in my mind, the point becomes an extension of the body, and the 'feel' at the point is translated into a TS-sensitised body site that demands even greater pressure until the point is broken.
>
> (Bliss, Sensory experiences of Gilles de la Tourette syndrome, 1980. p. 1347)

These intricate and complex experiences seem to strongly resonate by many people with Touretteic sensibilities, including the research participants of this study. There is, therefore, no lack of evidence that the circumstances under which compulsive interactions take place are crucial in their performance.

Nonetheless, in the scientific onto-epistemology, bodily surroundings are understood as inert and passive, and differences between places are treated as a given but have no ontological power. In fact, the context of compulsions is rather understood as an analytical nuisance that needs erasure from empirical research, hence the necessity of laboratory conditions under which neuropsychiatric, biomedical, and clinical knowledge is created. These laboratory conditions that underpin life scientific studies with a focus on frequency, which is often used to qualify improvement of a mode of interference such as behavioural therapies, render the person mostly static in a seated position facing a camera and/or observation booth in an otherwise unfamiliar, white, and mostly empty room. This includes studies that measure the impact of exercise on tic frequency, such as a study by Jackson et al. (2020) that asked children aged 10–12 to do Kick Boxing and Tai Chi exercises as mediated through an X-Box 360 Kinect, whilst wearing a heart rate monitor and with a camera pointed at them at two meters distance in a laboratory space. Similar to Morera Maiquez et al.'s (2020) study that tested the tic-reducing wrist band mentioned earlier, the analytical focus was only on simple tics in the face and upper body, which reinforces the hallmark position of these tics at the detriment of other movements, and, in the process, dismisses possibilities for a more holistic consideration of compulsions as subset of Tourettic motions.

Methodologies with positivist underpinnings do not only measure frequency in highly artificial circumstances but also forego more complex compulsions, including all interactions. These studies and other medical encounters thus cannot account for what happens outside the broader medical spaces. Nonetheless, the claims these studies make sustain the life scientific rationalisation of compulsive interactions, which can lead to confusions about certain situations, but can also directly oppose the experience of those diagnosed. The 'rebound effect' is one such ongoing dispute. It is the observation of people with Tourette's experiencing a strong increase of the need to do tics and compulsions after they 'suppress'[7] them for a period that cannot be registered following clinical methodology (Verdellen et al. 2008). People who perform compulsions almost unanimously disagree that it does not exist, as Dylan explains:

Dylan: Actually, everything you catch[8] speaking of tics, you get that. Scientifically, it is not proven yet; it's not confirmed that the rebound effect exists … But it exists. Why? Because it does.

DB: Well … yeah …

Dylan: That's why we don't need more supporting arguments, missus scientist!

Aside from a discrepancy between the two knowledge systems, it also highlights how compulsivity remains analytically entirely unaccounted for in the spaces beyond the direct gaze of the life sciences (see Beljaars 2020). The implications of this fourth transformation thus entail that the life sciences encourage to only consider compulsions as contextualised in quantitative measures, and through the way the person conceives of situations, rather than the situations themselves. This analytical conceptualisation reinforces clinical treatment as decontextualised as well; through biomedicine as well as through behavioural therapies. These push the message that compulsions are performed because the person cannot inhibit themselves, which transforms compulsivity into a matter of personal control in which the body is positioned as enemy which needs to be kept under control; regardless of the circumstances (see Hollenbeck 2003).

Moving forward

There are, broadly defined, four elements to the neuropsychiatric, biomedical, and clinical rationalisation of compulsive interaction that create a new set of confusions on top of those created in the frameworks that are currently available for making sense of compulsions (see Chapter 1). They entail discouragements to see differences in kind as well as in context of compulsion and encouragements to see compulsive interactions as always and only problematic, both from a personal and from socio-normative point of view. These four elements represent the transformations but do not provide complete answers to people's own experiences and conceptions of why they feel compelled to interact with their surroundings in a very particular way without wanting to, without knowing why, and often without anticipating they have to. Addressing them presents us with a requirement of a radically new ethics of analysing compulsivity.

The neuropsychiatric rationalisation of the determination that compulsive acts are a sign of a faulty brain leads to a reluctance to say which acts through the accepted idiosyncracy makes this rationalisation a dangerous one. It also takes away the possibility to express the idiosyncracy of compulsion. This rationalisation, however much currently felt as helpful answer by people with a diagnosis, is dangerous in that it effectively blocks the consideration of the compulsive phenomenon as performance. As such, we remain stuck in Lowri's frustration in which she cannot tell what movement she makes is compulsive. Indeed, rather than an understanding of what compulsion is, the biological narrative describes the boundaries between the 'healthy' and this version of the 'unhealthy' in conjunction with the 'abnormal' derivation of the 'normal' through the production of signs of subversion. Compulsion can never become considered on its own merit because it is precisely defined as pathological behaviour; the becoming abnormal; it is a colonisation of human movement and bodily expression. The life sciences can only ever locate the frontier and allow encroachment on unrationalised bodily performance. Indeed, we are reminded of Deleuze and Guattari's (2004 [1980]: 275) thoughts on the value of heterogeneity:

> The histories of ideas should never be continuous; it should be wary of resemblances, but also of descents of filiations; it should be content to

mark the thresholds through which an idea passes, the journeys it takes
that change its nature or object.

> (Deleuze & Guattari, *Thousand Plateaus. Capitalism and
> Schizophrenia*, 2004 [1980], p. 275)

Confusions, in part, emerge because of the immediate problematisation of
compulsions, which is the only explanation the neuropsychiatric sciences
offer, and their biologication confirms it. Hence all analytical effort is
focused on understanding this problem and then solving it. It does not formulate, nor question the conditions under which compulsive interactions
are understood as a problem; hence, it does not examine these acts beyond
their framing as a problem. The problem is a given that is decontextualised
beyond the body; in other words, the problem is emerging solely from the
person who needs help and seeks it through diagnostisation (one of the few
options available) not from the broader socio-cultural and political contexts.
Indeed, through diagnostisation that transforms a set of performances to a
problematic person, these people acquire a new ontic essence that sweeps
the person up in political processes of exclusion that makes them vulnerable to processes of dehumanisation and permits acts of colonisation; of
legal interference with the person through their body to make them more
'fully human' again. Therefore, an episteme that emancipates the individual
person and their circumstances by ontologically centring compulsive interactions can help mediate the vulnerability of these people to such processes.

Many participants discussed their struggles and difficulties with compulsivity and the part of their lives that they and other around them associ-
ate with Tourette syndrome. However, when discussing and performing the
individual acts – with some exceptions – the problematic dimension was
largely unimportant. Discussing compulsive interactions with Ginny and
her manners of coping with having to do them, she explains how completion
of some compulsions makes her feel:

> You know, if you were to remove it all, you'd become unhappy *laughs*
> That's true, you shouldn't want to remove it all with those tics and that,
> because you'd remove happiness. I really believe that, and you do have
> to make it work for yourself, but that you can really be in that moment of
> happiness! Yeah, you know, it doesn't have any function, and it doesn't
> have any content, but that just doesn't matter!

She was not alone in remarking on the liveliness of some of the experiences
emerging from aesthetically pleasing results of compulsive interactions.
Such slight and brief exhilaration might sprout from such an act precisely
because they are a-personal and unprecedented, and can thus retain an element of surprise to the person performing them. The latter might especially
be the case if the compulsion follows seamlessly on an other-than-compulsive
act. Nonetheless, Ginny and all other participants contend that having

to perform compulsive interactions remains stressful. This paradox high-lights the internal intricacies of compulsive interactions, which does not only produce experiential confusions for the person performing them. It also presents us with a slippage on favourability, functionality, and enjoy-ability, thereby conjuring up analytical confusions of compulsive interac-tions. In turn, they bring up questions of the justness of dismissing certain aspects of compulsive interactions in analysis. Indeed, rather than retaining analytical difference between compulsive and other-than-compulsive, or medicalised or not medicalised acts, these questions suggest that in order to understand compulsive interactions, they demand further individuation on both experiential and analytical levels.

Suspending the problematisation of compulsions does *not* encourage the rendering of these interactions as unproblematic; what it does is denote how the narrow definition of compulsion as a problem may have stifled further exploration and examination, and thus a broader understanding of the phe-nomenon. I argue that the confusions surrounding this phenomenon are at the heart of this, in particular the experiential ones, because they reflect the necessity of a thorough re-examination of explanatory power of the struc-tures that Western sciences have in place to understand human behaviour, and shed light on the blind spots they leave.

Notes

1 For lobotomy, these working include problems with keeping balance, walk-ing, swallowing, and speaking, as well as suffering from brain infections, spas-ticity and paralysis of all four limbs, and cognitive functioning (see Mukhida et al. 2008 for an overview). Other effects of DBS include seizures, problems with vision, and headaches (Testini et al. 2016, Marano et al. 2019), problems performing small movements (Huys et al. 2016), apathy, paraesthesia, erectile dysfunction, problems with emotions, weight gain, (Balderman et al. 2019).
2 This is a deliberate move away from psychoanalytical approaches to Tourette syndrome that had conceptualised them as *acts* to emphasise as a deliberately meaningful understanding of bodily action.
3 Original word: "donder" which is difficult to translate.
4 With thanks to Jo Bervoets.
5 Directed by Christopher Rawlence.
6 Some behavioural therapies, such as Habit Reversal Therapy, do differentiate between compulsions and tics to the extent that they target those acts that are most problematic for these people.
7 The possibility of tic suppression is currently (summer 2021) being challenged by many people with a Tourette syndrome diagnosis, as accurate description of what is experienced to happen. It suggests that not performing tics or com-pulsions – suppressing them – is the end of it. However, with the attempt to hold in tics, the pressure to perform them increases, so there might not be a 'net gain'. Also, many people experience having to perform many more tics after a period of holding them in, so instead of tics and compulsions disappearing, it constitutes a displacement over time. Therefore, indicating this phenomenon in the remainder of the book, the term appears in quotation marks.
8 Dylan uses 'catching' (original: 'opvangen') to indicate 'suppressing' the need to perform compulsions and consequently not performing them.

References

Alsobrook, J.P. II and Pauls, D.L. 2002. A factor analysis of tic symptoms in Gilles de la Tourette syndrome. *American Journal of Psychiatry* 159, pp. 291–296.

Asam, U. and Karrass, W. 1981. Gilles de la Tourette syndrome and psychosurgery. *Acta Paedopsychiatrica* 47, pp. 39–48.

Baldermann, J., Melzer, C., Zapf, A., Kohl, S., Timmermann, L., Tittgemeyer, M., Huys, D., Visser-Vandewalle, V., Kühn, A., Horn, A. and Kuhn, J., 2019. Connectivity profile predictive of effective deep brain stimulation in obsessive-compulsive disorder. *Biological Psychiatry* 85(9), pp. 735–743.

Bankey, R. 2004. The agoraphobic condition. *Cultural Geographies* 11, pp. 347–355.

Beljaars, D. 2020. Towards compulsive geographies. *Transactions of the Institute of British Geographers* 45, pp. 284–298.

Bervoets, J. forthcoming. Tourette Syndrome and Dynamic Moral Responsibility: A Healthier Look at Tourette's?. PhD Thesis. University of Antwerp.

Bervoets, J. and Beljaars, D. in review. From deficit to surplus models of mental illness. Tourette syndrome: A case study. *Social Science & Medicine*.

Bliss, J. 1980. Sensory experiences of Gilles de la Tourette syndrome. *Archives of General Psychiatry* 37, pp. 1343–1347.

Bloch, M.H. and Leckman, J.F. 2009. Clinical course of Tourette syndrome. *Journal of Psychosomatic Research* 67(6), pp. 497–501.

Bloch, M.H., Peterson, B.S., Scahill, L., Otka, J., Katsovich, L., Zhang, H. and Leckman, J.F. 2006. Adulthood outcome of tic and obsessive-compulsive symptom severity in children with Tourette syndrome. *Archives of Pediatrics and Adolescent Medicine* 160(1), pp. 65–69.

Capriotti, M.R. et al. 2013. Environmental factors as potential determinants of premonitory urge severity in youth with Tourette syndrome. *Journal of Obsessive-Compulsive and Related Disorders* 2, pp. 37–42.

Cath, D.C. et al. 1992. Mental play in Gilles de la Tourette's syndrome and obsessive-compulsive disorder. *British Journal of Psychiatry* 161, pp. 542–545.

Cath, D.C. et al. 2001. Repetitive behaviors in Tourette's syndrome and OCD with and without tics: What are the differences? *Psychiatry Research* 101, pp. 171–185.

Cavanna, A.E. and Nani, A. 2013. Tourette syndrome and consciousness of action. *Tremor and Other Hyperkinetic Movements* 3, pp. 1–8.

Christenson, G.A., Ristvedt, S.L. and Mackenzie, T.B. 1993. Identification of trichotillomania cue profiles. *Behavioral Research and Therapy* 31, pp. 315–320.

Clegg, J., Gillott, A. and Jones, J. 2013. Conceptual issues in neurodevelopmental disorders: lives out of sync. *Current Opinion in Psychiatry* 26, pp. 289–294.

Cohen, A.J. and Leckman, J.F. 1992. Sensory phenomena associated with Gilles de la Tourette's syndrome. *Journal of Clinical Psychiatry* 53, pp. 319–323.

Cohen, S., Leckman, J.F. and Bloch, M.H. 2013. Clinical assessment of Tourette syndrome and Tic disorders. *Neuroscience and Biobehavioral Reviews* 376, pp. 997–1007.

Conelea, C.A. and Woods, D.W. 2008. The influence of contextual factors on tic expression in Tourette syndrome: a review. *Journal of Psychosomatic Research* 65, pp. 487–496.

Conelea, C.A., Woods, D.W. and Zinner, S.H. 2011. Exploring the impact of chronic tic disorders on youth: results from the Tourette syndrome impact survey. *Child Psychiatry and Human Development* 42, pp. 219–242.

De la Tourette, G.G. 1885. Étude sur une affection nerveuse caractérisée par de l'Incoordination motrice accompagnée d'Écholalie et de coprolalie (Jumping, latah, myriachit). *Archives de Neurologie* 9(19–42), pp. 158–200.

Deleuze, G. and Guattari, F. 2004 [1980]. *A Thousand Plateaus. Capitalism and Schizophrenia.* London: Continuum.

Dowbiggin, I.R. 1991. *Inheriting Madness: Professionalization and Psychiatric Knowledge in Nineteenth Century France.* Berkeley: University of California Press.

Eapen, V., Mortiarty, J. and Robertson, M.M. 1994. Stimulus induced behaviours in Tourette's syndrome. *Journal of Neurology, Neurosurgery, and Psychiatry* 57, pp. 853–855.

Ferrao, Y.A., de Alvarenga, P.G., Hounie, A.G., de Mathis, M.A., de Rosario, M.C. and Miguel, E. 2013. The phenomenology of obsessive-compulsive symptoms in Tourette syndrome. In: Martino, D. and Leckman J.F., eds. *Tourette Syndrome.* Oxford, New York: Oxford University Press, pp. 50–73.

Goetz, C.G., Leurgans, S. and Chmura, T.A. 2001. Home alone: Methods to maximize tic expression for objective videotape assessments in Gilles de la Tourette syndrome. *Movement Disorders* 16(4), pp. 693–697.

Greenhough, B. 2011. Citizenship, care and companionship: Approaching geographies of health and bioscience. *Progress in Human Geography* 35(2), pp. 153–171.

Groth, C., Mol Debes, N., Rask, C.U., Lange, T. and Skov, L. 2017. Course of Tourette syndrome and comorbidities in a large prospective clinical study. *Journal of the American Academy of Child and Adolescent Psychiatry* 56(4), pp. 304–312.

Hollenbeck, P.J. 2003. A jangling journey: Life with Tourette syndrome. *Cerebrum* 53, pp. 47–61.

Huisman-van Dijk, H.M., Schoot, R., Rijkeboer, M.M., Mathews, C.A. and Cath, D.C. 2016. The relationship between tics, OC, ADHD and autism symptoms: A cross- disorder symptom analysis in Gilles de la Tourette syndrome patients and family-members. *Psychiatry Research* 237, pp. 138–146.

Huys, D., Bartsch, C., Koester, P., Lenartz, D., Maarouf, M., Daumann, J., Mai, J., Klosterkötter, J., Hunsche, S., Visser-Vandewalle, V., Woopen, C., Timmermann, L., Sturm, V. and Kuhn, J., 2016. Motor improvement and emotional stabilization in patients with Tourette Syndrome after deep brain stimulation of the ventral anterior and ventrolateral motor part of the thalamus. *Biological Psychiatry* 79(5), pp. 392–401.

Jackson, G.M., Nixon, E. and Jackson, S.R., 2020. Tic frequency and behavioural measures of cognitive control are improved in individuals with Tourette syndrome by aerobic exercise training. *Cortex* 129, pp. 188–198.

Kane, M.J. 1994. Premonitory urges as 'attentional tics' in Tourette's syndrome. *Journal of the American Academy of Child and Adolescent Psychiatry* 33, pp. 805–808.

Karp, B.I. and Hallett, M. 1996. Extracorporeal 'phantom' tics in Tourette's syndrome. *Neurology* 46, pp. 38–40.

Kimura, Y., Iijima, K., Takayama, Y., Yokosako, S., Kaneko, Y., Omori, M., Kaido, T., Kano, Y. and Iwasaki, M. 2021. Deep brain stimulation for refractory Tourette syndrome: Electrode position and clinical outcome. *Neurologia Medico-Chirurgica* 61(1), pp. 33–39.

Kushner, H.I. 1999. *A Cursing Brain? The Histories of Tourette Syndrome.* Cambridge, MA: Harvard University Press.

Kushner, H.I. 2008. History as a medical tool. *Lancet (London, England)* 371(9612), pp. 552–553.

Leckman, J.F., Riddle, M.A., Hardin, M.T., Ort, S.I., Swartz, K.L., Stevenson, J. and Cohen, D.J. 1989. The Yale global tic severity scale: Initial testing of a clinician-rated scale of tic severity. *Journal of the American Academy of Child and Adolescent Psychiatry* 28, pp. 566–573.

Leckman, J.F., Walker, D.A. and Cohen, D.J. 1993, Premonitory urges in Tourette's syndrome. *Journal of American Psychiatry* 150, pp. 98–102.

Lombroso, P.J. and Scahill, L. 2008. Tourette syndrome and obsessive-compulsive disorder. *Brain & Development* 30(4), pp. 231–237.

Mansueto, C.S. and Keuler, D.J. 2005. Tic or compulsion?: It's Tourettic OCD. *Behavior Modification* 29(5), pp. 784–799.

Marano, M., Migliore, S., Squitieri, F., Insola, A., Scarnati, E. and Mazzone, P., 2019. CM-Pf deep brain stimulation and the long term management of motor and psychiatric symptoms in a case of Tourette syndrome. *Journal of Clinical Neuroscience* 62, pp. 269–272.

Miltenberger, R.G., Fuqua, R.W. and Woods, D.W. 1998. Applying behavior analysis to clinical problems: Review and analysis of habit reversal. *Journal of Applied Behavior Analysis* 31, pp. 447–469.

Morera Maiquez, B. et al. 2020. Entraining movement-related brain oscillations to suppress tics in Tourette syndrome. *Current Biology* 30 (12), pp. 2334–2342.e3.

Mukhida K., Bishop M., Hong M., and Mendez I. 2008. Neurosurgical strategies for Gilles de la Tourette's syndrome. *Neuropsychiatric Disease and Treatment* 4(6), pp. 1111–1128

Neal, M. and Cavanna, A.E. 2013. Not just right experiences in patients with Tourette syndrome: Complex motor tics or compulsions? *Psychiatry Research* 210, pp. 559–563.

O'Connor, K.P., Gareau, D. and Blowers, G.H. 1994. Personal constructs associated with tics. *British Journal of Clinical Psychology* 33, pp. 151–158.

Palumbo, D. and Kurlan, R. 2007. Complex obsessive compulsive and impulsive symptoms in Tourette's syndrome. *Neuropsychiatric disease and treatment* 3(5), pp. 687–693.

Ramkiran, S., Heidemeyer, L., Gaebler, A., Shah, N.J. and Neuner, I. (2019) Alterations in basal ganglia-cerebello-thalamo-cortical connectivity and whole brain functional network topology in Tourette's syndrome. *NeuroImage: Clinical* 24, p. 101998

Robbins, T.W., Gillan, C.M., Smith, D.G., de Wit, S. and Ersche, K.D. 2012. Neurocognitive endophenotypes of impulsivity and compulsivity: towards dimensional psychiatry. *Trends in Cognitive Sciences* 16(1), pp. 81–91.

Robertson, M.M. and Cavanna, A.E., 2007. The Gilles de la Tourette syndrome: a principle component factor analytic study of a large pedigree. *Psychiatric Genetics* 17, pp. 143–152.

Robertson, M.M. et al. 2008. Principal components analysis of a large cohort with Tourette syndrome. *British Journal of Psychiatry* 193, pp. 31–36.

Rosario-Campos, M., Leckman, J., Mercadante, M., Shavitt, R., Prado, H., Sada, P., Zamignani, D. and Miguel, E. 2001. Adults with early-onset obsessive-compulsive disorder. *American Journal of Psychiatry* 158(11), pp. 1899–1903.

Rothenberger, A. and Roessner, V. 2019. Psychopharmacotherapy of obsessive-compulsive symptoms within the framework of Tourette syndrome. *Current Neuropharmacology* 17(8), pp. 703–709.

Sambrani, T., Jakubovski, E. and Müller-Vahl, K.R. 2016. New insights into clinical characteristics of Gilles de la Tourette syndrome: findings in 1032 patients from a single German center. *Frontiers in Neuroscience* 10, p. 415.

Sandle, R.V. 2012. The deconstruction of Gilles de la Tourette's syndrome. *Journal of European Psychology Students* 3(1), pp. 68–77.

Schroeder, T. 2005. Moral responsibility and Tourette syndrome. *Philosophy and Phenomenological Research* 71(1), pp. 106–123.

Shprecher, D. and Kurlan, R. 2009. The management of tics. *Movement Disorders* 24(1), pp. 15–24.

Silva, R.R. et al. 1995. Environmental factors and related fluctuation of symptoms in children and adolescents with Tourette's disorder. *Journal of Child Psychology and Psychiatry* 36, pp. 305–312.

Smeets, A. et al. 2018. Thalamic deep brain stimulation for refractory Tourette syndrome: clinical evidence for increasing disbalance of therapeutic effects and side effects at long-term follow-up. *Neuromodulation* 21(2), pp. 197–202.

Testini, P., Min, H., Bashir, A. and Lee, K., 2016. Deep brain stimulation for Tourette's syndrome: The case for targeting the thalamic centromedian–parafascicular complex. *Frontiers in Neurology* 7, pp. 193.

Turtle, L. and Robertson, M.M. 2008. Tics, twitches, tales: The experiences of Gilles de la Tourette's syndrome. *American Journal of Orthopsychiatry* 78(4), pp. 449–455.

van Bloss, N. 2006. *Busy Body. My Life With Tourette Syndrome*. London: Fusion Press.

van der Salm, S.M., Tijssen, M.A., Koelman, J.H. and van Rootselaar, A.F. 2012. The bereitschaftspotential in jerky movement disorders. *Journal of Neurology, Neurosurgery, and Psychiatry* 83(12), pp. 1162–1167.

Verdellen, C.W.J. 2007. *Exposure and Response Prevention in the treatment of Tourette syndrome*. PhD Thesis, Radboud University.

Verdellen, C.W.J. et al. 2008. Habituation of premonitory sensations during exposure and response prevention treatment in Tourette's syndrome. *Behavior Modification* 322, pp. 215–227.

Wang, Z. et al. 2011. The neural circuits that generate tics in Tourette syndrome. *The American Journal of Psychiatry* 168, pp. 1326–1337.

Wilensky, A. 1999. *Passing for Normal: A Memoir of Compulsion*. New York: Broadway Books.

Williams, M.T. et al. 2011. Myth of the pure obsessional type in obsessive–compulsive disorder. *Depression and Anxiety* 28(6), pp. 495–500.

Woods, D.W. et al. 2005. The premonitory urge for tics scale PUTS: Initial psychometric results and examination of the premonitory urge phenomenon in youths with tic disorders. *Journal of Developmental and Behavioral Pediatrics* 26, pp. 397–403.

Woods, D.W. et al. 2009. The development of stimulus control over tics: a potential explanation for contextually-based variability in the symptoms of Tourette syndrome. *Behaviour Research and Therapy* 47, pp. 41–47.

Worbe, Y. et al. 2010. Repetitive behaviours in patients with Gilles de la Tourette syndrome: tics, compulsions, or both? *PLoS ONE* 5, p. e12959.

Xu, W., Zhang, C., Deeb, W., Patel, B., Wu, Y., Voon, V., Okun, M.S. and Sun, B. 2020. Deep brain stimulation for Tourette's syndrome. *Translational Neurodegeneration* 9, p. 4.

3   Compulsive expressions

Having watched my sister's compulsions for years, and having observed the bodies of the research participants, I am keenly aware of them. In contrast to the calm bodily movements of the everyday, there is a certainty in compulsive movements. An eagerness almost. Bodies captured in compulsive moments are very confident bodies; they reach with calculated speed, push with exact pressure, tap, place, swipe with the dedicated precision of an Olympic athlete. Compulsive interactions stand revealed amidst the infinite possibilities of everyday life. They appear as immutable, they emerge like revelations.

To resurface from the confusions that are produced by and around compulsive performance, that escape from familiar cognitive behavioural structures and are only partially solved by neurobiological answers, compulsivity requires derationalisation. Therefore, this chapter further deconstructs compulsivity and sets it up to be rebuilt again. It does so by examining how compulsive interactions emerge in the flow of everyday life, and how this is experienced or how it escapes memory. In addition, it explores the variations in how they become known to people who perform them.

Awareness and memory

To return to the opening question of confusions and the partially useful answers the neuropsychiatric sciences can offer, we are left with the experiential dimension of compulsive interactions. Experience can provide us with an understanding as to what it is when we talk about compulsion, how it feels to be compulsive intermittently between further life. This section attends to what traces compulsive interactions can leave or how they go unnoticed.

For all participants, compulsivity was a distinctly demarcated aspect of their lives. At the beginning of the recording of her first mobile eye-tracking session, Elisa is fairly confident that she is aware of all her compulsive interactions when we watch the footage. She would drive us to her home, which was her idea because:

DOI: 10.4324/9781003109921-4

Elisa: That's when you see it most obviously

DB: Yes, you are aware of this

Elisa: Yes, yes, yes

DB: Many people who partake in this study are less aware of them

Elisa: Oh, I am aware of my compulsivity, I would rather not have that, it would be easier if you wouldn't have to do it

After further discussion and towards the end of the recording, Elisa reconsiders her certainty: "well, I think I wouldn't be able to recall *everything*, but a lot of things I actually would, yeah" (emphasis vocalised). Whilst the compulsive condition as the construct that names their medicalisation thus might be remarkably memorable; it seems to be less straightforward for the unfolding of individual compulsive interactions. Sage also struggles to list what kind of compulsive interactions she performs:

Like 'ah, I never do that!', and then you start thinking and then you're like 'ehhh, yeahhh'. That will start when Diana has left later on, and that I'm doing something and then I think 'ooh, I actually do!'

Sage's account similarly suggests that some compulsive interactions may only be recalled when residing in the same bodily situation, escaping memory outside these situations. Indeed, beyond those with particularly memorable emotional effects, such as humiliation, social awkwardness or pain, the compulsions participants elaborated on or demonstrated during formal and informal study meetings were often performed in the room we were in, and moving through the different spaces of their homes, cars, shops, etc., conjured up new ones. Watching their eye-tracking recordings allowed the research participants to revisit the actualised unfolding of the interactions, which poignantly demonstrated that, more often than not, they did more compulsions than that they were aware of doing at the time or could recall doing:

It's also funny that I can't, that you're not conscious of it. Talking about it is quite funny.

(Sage)

~~~

When you see this, you get all restless because of yourself. Then you think 'do you always do that, making these movements?'

(Nora)

Not entirely sure what to expect of her bodily appearance and compulsive interactions performed, watching the recordings with Nora likened a game of 'spotting' compulsive interactions. It seemed to reduce her anxieties as to what would be 'revealed'. As we were watching the recordings that feature her hands a lot, her husband was elsewhere in the living room. He overhears us when I think I see Nora do a compulsion:

DB: Was that something?

Nora: Well, I was looking at that as well ... I think I do this *glides her index and middle finger along each other*

DB: Yeah, for a moment

Nora: *Astonished* But I do that the whole time ... you *turns to husband* have said that before, that I do that constantly, but I don't know, gliding those fingers along each other

Husband: Yes, the whole time.

During the meeting to get to know each other and later on during the interview, she had proclaimed being a "secretive ticcer", explaining that she is very sensitive to the appearance of her bodily movements. She attempts to conceal those that could raise questions from others, so she states; "it surprises me because I do the same things all the time". Hence, she expresses or 'suppresses' her compulsions strongly based on her idea of others' sense of spectacle, which extends the clinical gaze of observability. In an informal conversation after the interview, she said she had been doing various kinds of compulsions outside my view under the table, including gliding her fingers along each other and her toes as well.

In addition to Nora being highly sensitive to particular compulsive acts, and remaining largely insensitive to others, those that escape her immediate awareness can include acts she does (very) often, whilst in other moments she remains strongly aware of doing them. Awareness of compulsive interactions or the lack thereof can therefore not solely be conceived of as a problem of memory. Likewise, Dylan and Sage speak of the shock of becoming aware of compulsive interactions *as they perform them*:

Some things thus go completely unnoticed (...) And the longer you actually talk about it, the more severe that tic comes back at you ... *looks down* ehm ... I'm stroking the desk with the back of my hand now! (...) I do lots of movements that I'm not even registering.

(Dylan)

~~~

Sage: Lately I've been more aware also, because then I'm practicing Exposure [and Response Prevention therapy] ... and then I am focussing so strongly on suppressing the [noninteractive] bodily tics that I ... – but then in the meantime, I'm doing something else because it helps to distract me. I do that for instance whilst I'm tidying the kitchen or something, and then I'm suddenly aware of all the things I touch!

DB: Just to touch?

Sage: *Because of the touching*

DB: Not because of//

Sage: //For real! You know, what I told you the other day – sorry – what I told you the other day, being focussed on particular ... shapes ehm ... but actually also when I clear the dish washer; everything I put in the cupboards I touch once more ... and that's how I notice more of them ... because when you're suppressing, you suddenly realise that that is also a tic! And I'm more aware now how irritating yeah, I'm more conscious of it, how, yeah how much I'm doing it, and that it's actually continual ...

A lack of awareness of performing compulsions might seem to juxtapose their experience of being very sensitive to their body's compulsive movements. Thus compulsions are performed regardless of people experiencing 'warning signs' like sensations that urge people to act, or conscious deliberation about the appropriateness of this particular compulsion in this situation. Diagnostisation and in particular behavioural therapies that focus on these signs, such as Sage's Exposure and Response Prevention therapy, as well as repressive social situations work to sensitise people to the upcoming compulsions which grants them time to decide to 'suppress' the urge or act out the required compulsion. Clinical studies reflect the presence and absence of urge sensations, with more 'complex' compulsions more likely being preceded by urges (Banaschewski et al. 2003, Verdellen 2007, Eapen and Robertson 2015). In addition to many other compulsions registered in this study, touching something one more time happens in the blink of an eye, and stroking the desk does not impede Dylan from discussing his compulsivity with me. However, rather than complexity measured in muscle group involvement, those compulsions that people are not aware of seem to go lack a relation to ongoing other-than-compulsive life. Whilst such invocation reminds of bodily automatism, Nora insists that compulsive interactions differ from automatic engagements:

DB: Oh! That was one, wasn't it?

Nora: *laughs* Yes! But one push too many!

DB: Because it should have been two?

Nora: Well, I had three. And that's not right

DB: Ok, and did you do something to counter [the effect]?

Nora: No, but I see this now, I think 'that's wrong, that's three *laughs* see?! Count them!'

DB: No, indeed, and in that moment you didn't think 'oh, it's one too many', you think?

Nora: I don't know. No, I don't know … and sometimes I just don't even want to give in to it, and then I think 'no, this is fine, there you go – done', you know?

DB: Yes ok, but then you do have to know in advance like 'oh I'll feel that proclivity to do it, or currently have that proclivity and I stop'?

Nora: Yes, but that's automatism, what I did there

While at first Nora recognises the movement of three pushes as compulsive, discussing it makes her recognise the third one as not compulsive but automatic. Bill also remarked on the ease with which compulsions can seemingly disappear in everyday motions: "you do it for such a long time, and what you said just now in the parking lot, lots of things happen in a flow, and now you really see it". Bill and Nora's accounts thus suggest that compulsions may seem to share the character of automatic action by *appearing* to be nondeliberate and, as Bissell (2011) invokes in the context of habit, 'unthought', when they are performed. However, further examining the experience depicts compulsive interactions as not quite automatic.

Looking at the recording with a mixture of curiosity, interest, and amazement, Bill watches himself compulsively organise a cutting board, cutlery, and glasses according to each other, and that has become part of the larger practice of clearing the dishwasher. These organising compulsions did not seem to break up his other-than-compulsive interactions:

[Bill: the compulsions] slipped in, I think eh … that if I don't do it that I, that I'd notice it … I think, yeah

DB: So these aren't things you necessarily sense you have to do?

Bill: No, I really do have to do them

DB: Yeah, so you can't just think 'oh, then I won't do them'?

Bill: No, I do really check, I saw myself properly watching like 'oh, turn that around, get the loop out'

(…)

DB: But it's not like you can do it in advance if you suspect you won't find it correct … that you can do it in anticipation if you will?

Bill: If I do it in advance … pfoo … no, that, that … no, I don't think so, no.

Despite Bill's prediction that he will be performing compulsions when he puts certain dishes in their normal place, he can neither *not* do them, nor pre-empt them. Hence, compulsions happen regardless of being recalled by the person performing them. Sage adds:

I think that generally speaking, it's so ingrained in my daily life that it's not annoying. (...) When you start noticing it more, then it becomes annoying all of a sudden.

In fact, Sage also offers an explanation as to how to better consider awareness of compulsive interactions. During one of the interviews regarding the mobile eye-tracking recordings, I ask her if it was just me seeing a certain, very short, and very easy to miss a 'lag' between her picking up an object and putting it in its new place. She ponders what happens:

Sage: I have that very often that when I touch something I feel I'm not treating it in a special way, but that I hold on to it just that little longer, or put just that little extra pressure on it, but that happens in such a fraction of a second that I often don't ... at least, maybe if you were to ask me in the moment. Then I'd know, but looking back at it now, then I think ...

DB: Yeah, so then it's not necessarily a choice, an active choice to do it?

Sage: No, more like [inaudible], more like 'I am laying my hand on the table now and I feel now ... very much the sensation of the table beneath my ... tip of my index finger' because with that one I feel all those pointed edges. So I sense that very explicitly, so I can experience that in the moment I just touch it briefly, so it's most definitely a tic because I feel that sensation, it is just preceded by a less eh ... elaborate performance or something, you get what I mean?

For Sage, the sensations of compulsive interactions cannot be ignored, but the interaction itself might only be realised upon unfolding, and might also remain 'in the moment'. Given the lack of contribution that a compulsion has to ongoing life, the irrelevance of such interactions reduces their capacity to form a memory beyond the compulsive 'requirement' or 'fulfilment' of the sensation. The lack of awareness of the performance of compulsive interactions does then not mean a disappearance of the body, which Drew Leder (1990) argues happens when focusing on functional life, when bodies are not in pain, and when they are 'moved by emotion' (Ricœur 1966). Nor does it mean a disappearance of movement of one's body during the act, like when contemplating during driving a car. Rather, how compulsions remain unremarked on seems mostly down to two things. First, it involves a disappearance of a capacity to relate to the bodily movement; and secondly, a lack of significant contribution to and interference with the other-than-compulsive 'flow' of everyday life, which refrains a memory of them to form. Alan elaborates on an occasion in which he became aware of and recalls performing a specific compulsion:

It [unknown object] is in the shed, this height, that's placed on a table or something, and at some point I'm having a drink and do this ... [taps his fingers on the table in the way he did on the object] touching all those small characters ... a few times back and forth (...) So at one point I think 'why am I doing that?' and the next day I did it again, so I've been doing that for a week now or something.

This description purports an interaction in which Alan almost seems to 'take part', as if he is made to interact and does not protest; he *registers* his hand reaching and his fingers tapping compulsively, as if he witnesses another body doing it. The situation demonstrates the almost unsurmountable separation and lack of communication between compulsive and other-than-compulsive subjectification processes. The 'de facto' character of his testimony of the compulsive performance remains entirely unquestioned, and Alan discusses his compulsive interaction on the same terms as his interaction with the glass that holds its drink. In Fleissner's (2007) terms, Alan describes his involvement in the compulsion as through being a *körper*; the biological and mechanical invocation of a body that needs sustenance, rather than *leib*; the lived and individuated body that relates to the self. As such, compulsive interactions can be experienced as a necessary evil, for which no negotiation is possible.

It follows that compulsivity has some sort of legacy, as it is actualised in compulsive interactions, highlighting the existence of a sensory kind of subjectivity that can make an impression, and is recalled when a purposeless and meaningless performance. This may explain Tomos, Siôn, and Mina's experiences of feeling 'out of it' and being overwhelmed by something that they do not consider part of themselves when they perform compulsive interactions. During these moments, the compulsions may be so intricate and elaborate that intended, other-than-compulsive life needs to be halted for the duration of the compulsion. Other-than-compulsive life then creates the situations upon which compulsive engagements can take place on a very momentary basis, whether they are performed in full awareness or escape it. This, in turn, has spatial connotations. Indeed, it may be possible to consider that in compulsive situations that remain unregistered, the body becomes 'hijacked' in its interactions with the bodily surroundings and further life is suspended for a moment.

By paying attention to the actualising bodily movements, we are beginning to see a more fully realised understanding of bodily action in which people become aware of a new set of interactions that had escaped their memories, bearing circumstantial reference to further intentional and moral life. This opens up a mode of spatial engagement beyond the familiar motions that can be analysed as purposive and meaningful, but that may have a strong impact on how spaces are embodied. This is true for people with the bodily disposition that have them diagnosed with Tourette syndrome, but given a lack of appropriate theorisations that consider bodily movement beyond purpose and meaning, this might also be true for a much broader population.

Knowing/feeling compulsivity

Knowing compulsivity and recalling it as sensation *as such*, and recognising the feeling after the moment of unfolding proves to be incredibly challenging. During the interview, Ginny struggles to put the experience of becoming compulsive in words:

> I can't analyse it properly now, but in the moment it's, you wouldn't notice, then, then you just do it ... kind of ... when I think about it now, that's how it kind of happens, that's how it feels.

For her and others, the feeling of compulsivity is very distinctive as it unfolds but also fades very quickly. Even the 'thrust' of being compelled into acting out an urge seems difficult to recall, as illustrated by Dylan when I ask him to explore how touching his nose compulsively feels different from other-than-compulsively scratching his nose: "I really just can't distinguish between ... I can't give a straight answer to that". Ginny's, Dylan's, and others' similar accounts echo the difficulties that Joseph Bliss' account (1980) testifies of. Bliss, a 62-year old person with lifelong experiences of compulsions as part of his Tourette syndrome diagnosis, published the first psychiatric article that mostly consisted of a first-person account of Tourette's, which became a key resource for further research on the condition. In it he sets out how compulsions (including tics) are unlike any other feeling or sensation:

> What is basic to the TS overt actions, and what differentiates them from normal bodily actions, is the (almost) intolerable need to produce a sharp punctuation that will at one and the same time gratify and terminate an *almost* intolerable urge. (...) [This punctuation sensation] can occur at varying degrees of intensity, from almost imperceptible, to soft and muted, (finally) to escalating, and in certain instances, to a high pitch of intensity. Which form they take depends on such factors as substitution activity in multiple tics, which tends to mute all actions; the attention-focussing effect in a single activity; and the stress of emotional situations.
>
> (Bliss, Sensory experiences of Gilles de la Tourette syndrome, 1980. p. 1345, emphasis original)

This detailed description of the sensations testifies complex internal variations. Aside from the characterising intensity or severity of the experience, compulsive interactions seem to remain largely nonrelational and constitute a sensory category on their own. Echoing Bissell in the context of pain, compulsivity "stubbornly *refuses* to be represented discursively" (2009: 911, emphasis original) and "resists being drawn into and subsumed by relation" (Dillon 2000: 5). Unable to be captured in language, it withstands comparisons with other corporeal states, such as hunger, thirst, or sexual drive. Whilst this prevents compulsion being considered through the explanatory structures that are built around these other states, and which would risk losing its characteristic nuances, it does present impossibilities as to how to understand and examine compulsivity. Nonetheless, in addition to Bliss' description, we can approach compulsive experiences from multiple angles.

Ginny explains that she recognises compulsivity in the added 'feeling' when she sees something. In the car, back from the supermarket where we did her second mobile eye tracking session, she explores the feeling as compelling element in full detail:

> Ginny: For me it often starts ... that I see something and that evokes a feeling. And then I want to do something with that feeling if it's a bad feeling, you want to order things or something, for instance. And if it's a good feeling, then you want to pick it up or do something with it, pinching it or something. That is, that's the compulsiveness, that it evokes a feeling. (...) In the shop I didn't really have that much compulsion, it's more that when I have more speed ... I don't feel it as much
>
> DB: yeah and of course I was around too.
>
> Ginny: Yeah but I did forget pretty quickly really ... but I have ... it differs a little bit per place, I think
>
> DB: why were you in a hurry now?
>
> Ginny: No but I am now used to doing it quickly I think, that that's why I ... eh ... felt less compulsive or something
>
> DB: and you allow yourself less time because of that?
>
> Ginny: Yes it makes a difference ... if you do it quicker, then you have, you don't have the thought, you know, then you don't have time for the thought. (...) and I think that I do certain compulsions also if I'm quick, but that these are just 'built in' or something.

Sage builds on the addition of this 'feeling' that can be 'bad' or 'good' to sensing a particular object as suddenly gaining a compelling quality; an 'intrigue' and 'strong appeal' beyond functionality, meaning, and beauty. For Sage, it is this additional intrigue and appeal that captures compulsivity. A few months prior to the interview, she and her partner had bought a deep fryer, and elements of this object became involved in compulsive interactions in particular situations: especially its capacity to become very hot:

> Sage: I really get ... batshit crazy because of that thing because I'm so incredibly intrigued by that hot frying fat
>
> DB: Uhuh
>
> Sage: I shall and must touch it. I <u>must</u>! Yeah ... you know it's superhot
>
> DB: Yes, painful and everything

Sage: But I shall and must know, and I can withstand it, but just being in the kitchen where we have the deep fryer is absolutely awful, and I think I rarely have had <u>anything</u> that has <u>such</u> strong appeal to me, and that really scares me because I'm afraid I might give in to it at some point.

Ginny, Sage, and others allude to the experience of compulsivity as an intensity articulating in everyday other-than-compulsive encounters and – potential – interactions. It is not necessarily the fryer, or frying fat itself, but Sage touching it that is intriguing. She does not feel forced to touch the hot frying fat because she would want to burn herself, nor is the intrigue related to a threatening thought. Sage's description of feeling so intensely compelled to interact with an object resonates strongly with the description of experiences other participants provided. Indeed, the intensity distinctly differs from how compulsivity is experienced in relation to obsession, as Sage concurs:

I do have compulsions [vis-à-vis obsessions][1], but these ones are more like compulsive acts, so more like, what you do with your research, with objects and all, but it's not driven by fear anymore. It really has to do with ... The bodily sensation, like 'hey, I like touching that', or 'I have to touch that', and not anymore like 'if I don't do that, then ...'

Both Ginny seeing something and interacting with it to prior to the thought and Sage's description of intrigue, appeal as sensation depict compulsivity as knowable through the general or 'empty' sensory anxiety it conjures. This is decidedly different from Sage's invocation of the absence of fear, and Ginny's invocation of a looming feeling. The absence of a certain "content" of a quality that can be named makes compulsivity a pre-personal phenomenon. In compulsivity based on fear or obsessions, which characterises compulsions in Obsessive Compulsive Disorder (OCD), the interactions utilise objects to negotiate the severe meaningful connotations. For instance, 'if I don't tap the railing of the stairs right, someone I know will die' or 'if my shoes I wore outside touch the living room carpet, I will get very ill'. Such compulsions are always excessive in the ways in which objects become involved, either in symbolic representational sense or without any reference to the content of the fear or obsession. Whereas the obsession's ideational legacies outlive the acts, and the objects are only involved to mediate the anxiety that comes with the fear or obsession, the intensity of the compulsions examined here does not precipitate into other-than-compulsive life occurring before, alongside and after it, apart from its sensory implications, such as potential pain from interacting. In these compulsions, the interactions are about these objects and their particular features. The 'empty' kind of compulsivity incites a deepening of mundane experiences with a strong affective resonance with the object that becomes involved in the interaction.

Besides failed sense-making exercises, Sage's invocation of 'appeal', Ginny's distinctive 'added feeling', and Bliss' description of 'punctuation' emphasise an unpredictability, and as such, an unknowability of compulsivity. How else, then, can we *know* or *feel* compulsivity? Perhaps it is precisely the ways in which some compulsions fail to be captured in an essential sort of way is also the reason why they can, but in a different mode. The unrelatedness to the flow of everyday life expresses in various forms of disruption, as the most memorable ones are often accompanied by intense feelings of intrigue and rightness that stand out in their severity.

Disruption and anticipation

In addition to recalling performing compulsive interactions and their distinct quality, they also gain presence through the timing of the requirement to perform them: compulsivity expresses in a disruptive mode. Interrogation of compulsions through this mode provides new insights into the circumstances of their emergence as well as on how their performance can be negotiated. Despite being performed by the person themselves, these interactions can thoroughly derail other-than-compulsive life in several ways.

Compulsivity expresses most violently when the person has no or very little time to adjust. To mind comes the example of Alan, who notices his hand reaching and his fingers tapping the unknown object in his garage. Another is performed by Dylan, who understands his compulsive tendencies mainly on their disruptive dimension. He felt compelled to repeat certain interactions, because the first time he did them felt wrong. Disruptions caused by compulsivity express with him acquiring the right sensations:

> I have this idea that ehm … Tourette's often disturbs the processing of a signal, and that this means that it has to happen again. (…) And because it's not right in your neurotransmitters, the signal has to go again and again. (…) sometimes a signal comes from somewhere else; you bump your elbow into a door, because you walk through it with your hands full. You touch the door, but then there's a stimulus you hadn't caused yourself, or something. Ehm … and that means that it has to happen again, otherwise the stimulus doesn't dissolve.

During our meetings, he regularly had to repeat interactions that are intended and part of planned activities but were not performed correctly. They stopped him in his tracks on multiple occasions, and interrupted his 'flow' of intentional acts, which in his words are an "overlap of the normal and abnormal". Every time Dylan *was involuntarily encountered by* object matter[2], it incited violent compulsivity.

Some interactions have a violently disruptive urge 'thrust', as they have a strongly situated character; Elisa's compulsive requirement to look at lampposts whilst driving for instance. Feeling forced to look at them, these interactions could not be postponed, hence driving on streets and junctions with many lampposts thus required her to interrupt her driving activities for longer.

Additionally, in a situation in which she felt she had 'missed' one, she would have to turn her head backwards to find it and look at it. Driving thus becomes a continually interrupted activity that considered in Michel Serres' (1995) terms, puts Elisa in an almost continual state of 'turbulence'. Turbulence, as an intermediary state between order and disorder, draws attention to her switching between and adjusting to driving and compulsively interacting, as being compelled to look at the lampposts could not be postponed or successfully 'suppressed'. Turbulence then names the experiential process of the compulsive disruption and provides an indication of the exhaustion that it produces. The switch between driving and compulsively looking does not only take away much needed attention to the driving practice, it ruptures Elisa's subjectivity in her sense of self and trail of thought, every time she has to look at a lamppost.

In Elisa's compulsive episode, the lampposts cannot be avoided to be looked at and are crucial elements to both compulsive and other-than-compulsive life. That implies that in their capacity to disrupt within compulsive moments, these objects have a certain multiplicity. Such object multiplicity also appears in the small objects that Sage cleans during her second mobile eye-tracking session. Dusting one small object after the other on display presented her with different objects quickly passing through her hands: she picks up a statue of a Greek god and dusts it, then goes to put it back in its place. Suddenly her hand holding the statue stops mid-air, and her right index finger forcefully presses in the hollow of its back. Rubbing it a good few times, her hand brings it back to 'its place', because it distinctly has one, she would clarify later on. With the object already in her hand, she has very little to no time to become aware of an urge and decide whether or not to 'suppress' this urge, which gives the compulsive act to follow a strong sense of immediacy. Sage argues that compulsions need to be performed as soon as possible, to avoid what she calls "stagnation in your normal activities", which would also reduce being caught in a state of turbulence.

Despite Siôn having a relatively long time – multiple seconds – to become aware of the pending compulsion, and to then decide if he 'gives in' to the urge to compulsively reorder the cushions on his sofa, the compulsion is still very disruptive. Setting the table for lunch in his living room during his third eye-tracking session, he had to repeat this twice more. He elaborates:

> Siôn: It's not that I decided to do it an hour in advance, like 'this isn't good, I still need to do it', It really just pops up in my head or something. I'm repositioning these things now
>
> DB: Whilst you hadn't entirely finished setting the table.
>
> Siôn: No, indeed, certainly not!

All tasks in his living room are therefore often interrupted, which causes him to sometimes forget what it was he was doing. He hesitates where to go next and this gaze switches between the kitchen and the lunch table, before he continues setting the table. Also Lowri explains that she often has to put everyday activities on hold because an engagement with particular objects could become problematic as it could suddenly turn compulsive:

It's more like from getting up [in the morning] until going to bed, and if by chance I see something, ah … then I have to touch that for a second if I, if I'd have to … or I'd have to, like, I'm putting the laundry up, ehm … I pick up two clothes pegs, then I have difficulties with choosing these pegs. There's always something, like, that makes me get stuck continually. I always have to, like, interrupt, like, hold back. Then you have to get over that barrier.

'Getting stuck' causes her to always have to prepare for other-than-compulsive activities to take extra time, because she finds it difficult to 'get over that barrier' and continue the activity she had been doing before. Objects that are interacted with in other-than-compulsive ways will thus always have the potential to incite compulsive interaction. Whereas compulsive interactions present as severe transgressions of normative and social, considering them in an anticipative, disruptive mode suggests how they also pose fundamental challenges to an individual's expectations about their body's role in expressing themselves. In addition, compulsivity purports another way of highlighting how all objects that pass people's hands have 'complex presence' (Mol 2002). This holds that interacting with objects regardless of their level of familiarity will always affect bodies in ways beyond comprehension.

Compulsive interactions can be anticipated to a certain extent. For Bill, leaving the house requires a series of compulsive interactions that have to be performed before he can step outside:

It happens throughout the day, that's why it means that they are kind of short interruptions in movements he makes, because he walks for instance through the door to his living room. Well, he has to touch the doorknob in a certain way and kind of tap it for a bit when he walks away.

(field notes[3])

In addition to compulsively interacting with all knobs or doors he passes through, he has to take one puff from a small cigar in the kitchen, he has to completely open and close the dishwasher upon walking past it to also compulsively check if the backdoor is closed. The disruptions continue in the supermarket where he has to carefully position and pinch all items for a few seconds before putting them in his basket. The full engagement of his body with these compulsions prevents him from continuing walking through the supermarket, looking at other items on the shelves, or checking his grocery list. These everyday practices are thus interlaced with strictly regularly occurring and similar, often recurring compulsions. Mina elaborates on similar experiences of having to deal with acting compulsively that she knows she has to go through with before it happens. She compares it with elements of her everyday life that are also habitual:

> A ritual of waking up in the morning, drinking coffee, checking up on what happened is not really a ... That's more like a habit. (...) Rituals are more innocent (...) but, for example absolutely having to sit at the right side in the train is a ritual, but with a mandatory character.

As she knows that these 'mandatory rituals' would have to be obeyed, she could prepare for becoming compulsive and slipping into performing a compulsive act. Such interactions therefore do not carry the shock of immediacy but are disruptive both in terms of the strict demands it poses to everyday tasks they are intertwined with, and in their capacity to produce major anxieties when these demands cannot be met. As the train poses a challenge for Mina, for Joe, and others, museums are places that are similarly always severely disruptive in anticipation and actuality. Joe reflects:

> But, you know, it's just what, eh ... – where you are and with whom you are. Look, for instance, when I walk in a museum or wherever, then I'd just like to handle everything.

Museum environments, trains, and other spaces are also disruptive for the 'suppressions' passing through them require; in the train when Mina cannot sit on the right side and museum spaces with displays of art that Joe can reach to touch, regardless of being allowed to do so. As will be considered in greater detail in Chapter 4, 'suppressing' the urge to compulsively interact is very tiring and refrains the person from doing and enjoying ongoing life in these spaces. For individuals, such spaces agglomerate into a geography of more or less certain disruption created by stringent compulsive needs and requirements.

The kinds of compulsions Mina and Joe describe, constitute situations that these people are compelled to produce. In other words; having to sit on the right side of the train could be seen as not causing a disruption of travelling by train, but *a situation that needs to occur*, otherwise the train trip may disrupt the experience of further life as it *should* unfold. The sensations that build up to the compulsion could then be conceived of as the turbulence that is experienced before a situation is 'corrected'. As such, compulsive interactions act as kind of immanent direction of or to the transcendental unfolding of life (after Rose 2006, Wylie 2009) brought about through a turbulent concoction of the shocks of deterrence, friction, and feverish seduction that pull the body towards, and push it away from the material surroundings. As this works on a pre-personal level, it is easy to see how having to fight this turbulence is exhausting.

Managing the disruption incited by compulsive interactions is possible by employing multiple strategies. Less engaged and elaborate movements leave possibilities for uninvolved body parts to act compulsively. In Sage's other-than-compulsive act of moving a candle holder, some of her fingers retain the possibility to engage with it compulsively, and the mobile eye-tracking recordings show her thumb pushed between the rim and the bulge of the handle. In part, avoiding potential disruption works through recognising when which compulsions could have to be performed and planning for them to happen in accordance with bodily capabilities, object capacities and the organisation of the space the compulsion may have to be performed. Such an intricate organisation becomes visible when Sage walks holding a mug to the kitchen and manages to finish compulsively pushing her thumb on its rim before putting it down on the counter. In another moment, the compulsion could require Sage to hold on to the mug to finish the compulsion whilst already standing in front of the counter or slow her pace to delay arriving at the counter.

Managing the potential disruptions of compulsive performance thus evokeds a wide variety of strategies, of which the full range is displayed throughout the chapters that follow. Some strategies help not just with particular compulsive interactions but help minimise potential disruptions. As Ginny argued, moving quickly can be a successful strategy. This was a strategy Joe used as well: anticipating he would have to touch chestnuts, the wool fur of sheep and tree bark during the observation, his pace was high. He also had his hands in his pockets and mostly looked down to the ground in front of him. Keeping the visual perceptions of his surroundings to a minimum and restricting the movement possibilities of his hands and arms kept the disruptions to a minimum, as it prevented him noticing and quickly reaching out to nearby chestnuts, sheep, and tree bark. The different elements and dimensions of the disruptiveness of compulsive interactions thus highlight a multiplicity of quickly shifting micro-geographies that emerge with the body and its surroundings.

We are beginning to see that despite being performed by individuals, compulsions destabilise the idea that people can always recall what their body does and prevent it from doing something unwanted in combination with the destabilisation of a continuous subject. We are seeing a disaggregation of the individual; compulsive interactions demonstrate that it is possible to disturb an individual to their core. Not just in terms of meaning, purpose, and morality, but in terms of self-reflection, bodily control – we see a crisis in the individual's capacity to express themselves through their body – basic intentionality, as well as in terms of their sensibilities. Perhaps that means that we can see how compulsivity names a crisis of the human figure in liberalism. Examining the pieces of the deconstruction this chapter brought about, we will explore the ethical repercussions of the singular situations of compulsive moments, in the various kinds of absences of the human figure. The next chapter considers how compulsions are constituted, dissects how the compulsive processes crumble human agency, and how the relations between bodies and their material surroundings need a radical rethinking.

Notes

1 The Dutch language, recognisant of the German expressions as coined by Sigmund Freud, has two words for compulsion; 'dwang' and 'drang'. Dwang alludes to compulsions following obsessive thoughts that connote fear, and drang alludes to compulsions following urges without an emotional aspect; these connote bodily movement incited by sensations, and/or feeling inexplicably compelled to make a particular movement.
2 Having established that compulsive interactions are not performed by the human with less human dominance as in other-than-compulsive engagements, I use a more passive language in the text to indicate the diminished human involvement in compulsive performativity.
3 The voice recorder had failed to record the interview with Bill, and I recorded as many things he had said as I could immediately after it took place.

References

Banaschewski, T., Woerner, W. and Rothenberger, A. 2003. Premonitory sensory phenomena and suppressibility of tics in Tourette syndrome: Developmental aspects in children and adolescents. *Developmental Medicine and Child Neurology* 45, pp. 700–703.

Bissell, D. 2009. Obdurate pains, transient intensities: affect and the chronically pained body. *Environment and Planning A* 41(4), pp. 911–928.

Bissell, D. 2011. Thinking habits for uncertain objects, *Environment and Planning A* 43, pp. 2649–2665.

Bliss, J. 1980. Sensory experiences of Gilles de la Tourette syndrome. *Archives of General Psychiatry* 37, pp. 1343–1347.

Deleuze, G. and Guattari, F. 2004 [1980]. *A Thousand Plateaus. Capitalism and Schizophrenia*. London: Continuum.

Dillon, M. 2000. Poststructuralism, complexity and poetics. *Theory, Culture and Society* 17(5), pp. 1–26.

Eapen, V. and Robertson, M.M. 2015. Are there distinct subtypes in Tourette syndrome? Pure-Tourette syndrome versus Tourette syndrome-plus, and simple versus complex tics. *Neuropsychiatric Disease and Treatment* 11, pp. 1431–1436.

Fleissner, J. 2007. Obsessional modernity: the 'institutionalisation of doubt'. *Critical Inquiry* 34(1), pp. 106–134.

Leder, D. 1990. *The Absent Body*. Chicago: University of Chicago Press.

Mol, A. 2002. *The Body Multiple: Ontology in Medical Practice*. Durham, NC: Duke University Press.

Ricœur, P. 1966. *Motivation and the corporeal involuntary*. In: Ricoeur, P. ed. *Freedom and Nature*. Evanston, IL: Northwestern University Press.

Rose, M. 2006. Gathering 'dreams of presence': A project for the cultural landscape. *Environment and Planning D* 24, pp. 537–554.

Serres, M. 1995. *Genesis*, trans. James, G. and Nielson, J., Ann Arbor, MI: Michigan University Press.

Verdellen, C.W.J. 2007. *Exposure and Response Prevention in the treatment of Tourette syndrome*. PhD Thesis, Radboud University.

Wylie, J. 2009. Landscape, absence and the geographies of love. *Transactions of the Institute of British Geographers* 34, pp. 275–289.

4 Urgency

On becoming compulsive

> To see something is to see what it is for; we see not shapes but possibilities.
> (Alphonso Lingis, 1995, *The World as a Whole*, p. 602)

By virtue of their capacity to create a momentary rupture, even when anticipated and lived through countless times before, compulsive interactions are events; revelatory and therein perhaps revolutionary. Their emergence requires a ceasing of purposeful, meaningful, and moral life, as if such life blinks for a second and everything disappears: the incessant grip held on the anticipated immanent future lost. With the termination of the punctuation, the achievement of just-rightness, and order reinstated, life becomes reborn and reorganises to its state in which the compulsion demanded it to stop. After such rebirth, all is familiar but new nonetheless. Thoughts need to be rethought, everything in sight needs to be re-seen, and tasks need to be re-intended. The absence of rationalisable incitements to move one's body in compulsive ways, in combination with a seemingly limitless multitude of interactions, compulsivity is embedded in, and emergent from, the pre-individual sensory realm. This chapter explores how this sensory realm forms the foundations for compulsivity, and in particular, its spatial underpinnings.

For people with compulsive experiences, compulsive action is inescapably tied to, and emerges from, a state of awareness that Michael Kane experiences as 'hyperattention' (1994), and Crossley and Cavanna (2013) describe as 'hypervigilance'. Joseph Bliss (1980: 1345) simply calls it "the Tourette state". Siôn explains that living this state means "that [he is] always aware of what happens around [him]" and no mediation of this awareness is possible. He elaborates: "your senses are constantly open", which incites him to continually "scan the room" to notice, or almost study, every detail of a room. Sara, comparing her experiences to her understanding of how familiar others perceive their environment, adds that for her "it's all more intense". Elisa echoes her and Siôn in arguing "what comes in does so way too powerfully". This unchosen 'bodily openness' to the world creates the

DOI: 10.4324/9781003109921-5

foundations for one's entire sensory life, and as such presents a real vulnerability, as Ginny explains:

> I'm just hyper-sensitive, it's a complete package (...) I respond violently to everything, you know. 'Too much'. So my controls are just wrongly adjusted there.

This hypervigilance makes Bliss (1980: 1345) feel "pathologically itchy", as it expresses as unrelenting sensations. For instance, it makes people feel the label of a sweater in their neck throughout the day, and/or requires one to continually look at a distortion in the wallpaper, and/or demand that one constantly looks for or counts similar or grouped objects (see Kurlan et al. 1989, Cohen and Leckman 1992, Karp and Hallett 1996). For instance, Nora explains that when she can see it, she counts the five flowers on an ornament that hangs on her living room wall, and in addition to clothes' labels, Sara severely dislikes feeling the fabric of her jeans on her skin:

> I have my top in my jeans, and to avoid feeling my jeans I have another shirt directly on my skin, so that it has something in between, you see. And when I put my hand in my pocket, I do feel on my leg that that pocket is there, but that's not too bad, because it's a soft and thin fabric. But otherwise it's really very annoying.

Similar to experiences voiced by autistic people, this hypervigilance can quickly be overwhelming, which is confusing, as sensory information does not translate into meaning, and physically uncomfortable or even painful (see Gerland 2003, Williams 2005). Participants in Belluscio et al.'s study (2011) contended that flashing and fluorescent lights and rapid colour changes can demand so much attention for a prolonged amount of time that seeing anything else becomes very difficult. Peter Hollenbeck (2003) elaborates on his careful management of his surroundings to avoid becoming overwhelmed:

> Bright, shifting lights against a dark background exacerbate many of my other motor tics, so I go to great lengths to avoid driving at night. When this event is unavoidable, I stay off main streets where the noxious stimuli of streetlights and headlights await me.
>
> (np) (Hollenbeck, A jangling journey:
> Life with Tourette syndrome, 2003, np)

In addition to remaining aware of being touched by a piece of clothing, the hyperawareness also expresses in people being bothered by light touches and preferring to be held firmly. Some participants to Belluscio et al.'s (2011) study even attest to purposefully engaging in activities that can be painful, such as pushing the skin back from under a nail), which reminds of Bliss' (1980) description of compulsions ending with some form of 'punctuation'.

These engagements seem to be a reciprocation or a mirroring in which the perceived harshness of the environment is met with an equally harsh engagement that resembles this harshness and punctuation. In other words, the surroundings meet an individual in this register, therefore, this person needs to meet their surroundings in this register as well.

Whilst most participants argue that their bodily environment often overwhelms them, particular bodily states are experienced to exacerbate this openness. For instance, for Ginny, not getting enough sleep or not sleeping well is associated with feeling "even more alert". "Then [her] shutters have opened even wider", and "it seems that [she] receives even more stimuli, or something ... yeah, then they just come in more forcefully". Additionally, when she starts menstruating, she feels more aware of and more easily bothered by the organisation of the furniture in her house. In particular the living room becomes a turbulent space and she is compelled to change it. Furthermore, Bill's experience with having chronic pains makes him feel particularly easily affected on days when the pain is more intense.

Neurobiological and neurocognitive research cannot register these experiences in the sensory systems and studies agree in their conclusion that people with a Tourette's diagnosis do not have an 'enhanced peripheral detection in the sensory tissues' (Hollenbeck 2001, Leckman et al. 2006, Belluscio et al. 2011, Cavanna and Nani 2013, Cohen et al. 2013, Sutherland Owens et al. 2013, Houghton et al. 2014, Schunke et al. 2016). Rather than in the sensory systems, another strand of life scientific research considers hypervigilance to be rooted in 'altered sensorimotor processing systems' in the brain, which alludes to differences in the brain's recognition of information that has been conceptualised as 'relevant' for the individual's survival (Belluscio et al. 2011, Cox et al. 2018). It underpins experiences of hypervigilance as an intense and mutual amplification of perception and the bodily action (Abbruzzese and Berardelli 2003, Leckman et al. 2006, Beste and Münchau 2018, Petruo et al. 2018). Such research begins to describe, but cannot quite capture, the excessiveness of hyperattentive experiences like those from Nick van Bloss (2006). When he enters a room he instantly *knows*, rather than *actively counts*, the number of chairs. He cannot avoid perceiving an overwhelming amount of detail and to avoid feeling compelled to count the chairs he actively searches for distraction. Environments thus present as a strongly provocative world, as Ginny attests:

Ginny: For me it often starts with that I see something and that it evokes a feeling. And then I want to do something with that feeling if it's a bad feeling, you want to order things or something, for instance. And if it's a good feeling, you want to pick it up or do something with it, pinching it or something.

~

Ginny: It feels like you're head over heels in love, that's how it feels, [Expression of frustration]. When you're in love you also feel like

> [Expression of excitement] (...) I find that such a terrible feeling! Yes! (...) You can be in love and wholly enjoy it or something, but it's also ... If I only just think about it, I'm getting itchy, you know, like [Expression of excitement]
>
> DB: You become anxious
>
> Ginny: Yes, then that just gives me such bad itches all over, that comes out of my ears! (...) like someone's touching your ears like that *tickles her ears and shudders*, or like with your nails scraping a chalkboard, like [Expression of excitement]

Ginny using 'wanting' in the first quotation alludes to a noted a-personal need within the context of 'doing or ordering something' after seeing something, and does not extend into further personal life. The vast majority of the research participants used such phrasing to elaborate on becoming compelled to act compulsively. As Ginny's second quotation attests, remaining in a constant state of anticipation to be provoked into interacting with one's bodily surroundings in ways that cannot be rationalised produces not only an existential openness to the world that cannot be stifled; an unrelenting exposure to the seductive forces of all that is present. It also (re)produces one's body as a sensual, almost magnetic entity that is pulled to move in accordance with the attractions of these surroundings.

In turn, this hyperawareness (re)produces the surroundings as landscape of "joined intensities: different and often ungraspable points of connection and cluster", to quote Bissell's work on his spatial constellation of pain (2009), in which he draws on Deleuze's (1991 [1988]: 92) conceptualisation of intensity as that what "gives all the qualities with which we make experience". Compulsivity reveals the bodily surroundings as presenting a multiplicity of intensities that create an ongoing, changeable, and fluid field of tension that bodily capabilities to respond are mapped onto. The incessant hyperattentive or hypervigilant sensibilities create the sensory conditions upon which compulsions can arise, and in turn, the bodily openness is a prerequisite for compulsivity. It is not *if* compulsions take place, rather *when and where*, which is a function of the correspondence between the environment and the body as mediated through perception.

Acceleration/urgency

If compulsivity emerges from a landscape of joined intensities produced and reproduced by the body and its surroundings as fuelled through perceptions directed by hypervigilance, the question is what exactly creates the acceleration that urges the performance of a compulsion? As not all compulsions are memorable beyond the act (see Chapter 3), the urge sensations are the experiential indicator of the acceleration of the necessity of the body and surroundings coming together in a compulsive interaction. In the clinical sciences, these sensations have been conceptualised as 'premonitory urges'. All participants echo Bliss (1980: 1344) in his assertion that "[t]here is really no adequate description of the sensations that signal the onset of the actions". Therefore, descriptions have been made using a plethora of metaphors with Joseph Bliss and Lance Turtle providing examples:

> When highly activated, the sensation immediately preceding a movement can be compared, inexactly, with a number of other types of feelings: (1) a compelling, subtle and fleeting, itch; (2) the moments before a sneeze explodes; (3) the tantalising touch of a feather.
> (Bliss, Sensory experiences of Gilles de la Tourette syndrome, 1980. p. 1344)

> The sensation is that of something welling up inside my head, or rather the premonition of something welling up inside my head, starting in the center, moving upward, and then seeming to then fly out of the top. The feeling will only fly out of the top of my head if I tic; this represents completion of the sensation and relief from it.
> (Turtle & Robertson, Tics, twitches, tales: The experiences of Gilles de la Tourette's syndrome, 2008, p. 451)

Paraphrasing Turtle's description, Sara describes it as an "itch that increases in intensity with time passing". Dylan explains that it could feel "as if there's something wrong in your nervous system", and that it can even be painful when he does not "give in" to the urge sensations, when the "pressure builds" in the body part compelled to take part in a compulsive act. Sage elaborates on the way she experienced the acceleration:

> Yeah, so Yesterday suddenly with that coat of [partner's name], I thought 'ooh, that's interesting!' (...) so now it's indeed more so that if my eye falls on something, like that element of his coat[1] that I think 'ooh, I really have to touch that for a second'.

Sage expressing "that [she] actually can't not do [the compulsion]", indicates the sensation as an escalation from "just being interested" to "feeling compelled to do a real touch tic", echoing experiences of other participants in this study, autobiographical writers (e.g. Kane 1994), and participants in clinical research. These sensations become more uncomfortable when the performance of the compulsion is delayed, or when told to stop or being teased by others (Leckman et al. 1993, Woods et al. 2005, Conelea and Woods 2008, Conelea et al. 2011, Wang et al. 2011, Capriotti et al. 2013). In this case, the distinction articulates the acceleration towards compulsive action more in terms of intensity and less as a particular quality (see also Cavanna et al. 2017). The urge comes into being when Sage's gaze captures the coat element; not the other way around. That is, the object does not become enrolled in a compulsive interaction because the person experiences the compelling sensations and the object happened to be nearby. Quite the opposite seems to happen; the bodily openness forms a bodily capacity to encounter the surroundings in ways that evade traditional rationalisation of bodily action. The urge sensation does not forego the object and it thus co-constitutes the sensations. Sara illustrates this in her explanation why she often holds small stuffed animals close by:

> Sara: I really have to have something in my hands or feel something, you know, and it's always one of those stuffed animals because other textures don't work and I get a bad feeling if I try to hold it, or I get irritated
>
> DB: Okay, and is that because of the same kind of structure?
>
> Sara: No, it's kind of on the inside – that you can roll the seam between your fingers, and this is just – this was my first, that on the top of it it's smooth but that's only with my middle finger. I don't have that with other fingers. (...) If I'd have my hands off them now, or something, I don't feel the tingling, but if I'd lie down for a while, I would want to pick them up again.

The tingling urge sensation locates in her middle finger, and by picking up the small stuffed animals, it disappears. Between her hands and two small stuffed animals, an ongoing spatial dialectic is formed. Despite the urge not specifically pointing out how the body should interact with the object, the urge does qualify in the direction of the object matter to interact with, in addition to the body parts that would be involved. The sensation of the acceleration is thus invariably responsive to the object and the involvement of the object is not a coincidence. The bond between her fingers and the stuffed animals remains actively felt outside the interactions. When she sits at her desk in the living room, the stuffed animals lie next to her chair on the floor, and when travelling by train she rolls them between her fingers or keeps them in a special compartment in her bag.

Whereas participants in Kwak et al.'s study (2003) confirmed urge sensations to emerge in the 'face/head', 'shoulders', 'arms', 'hands', 'neck', 'throat', 'feet', 'abdomen/stomach', and 'other', which was confirmed by the participants to this study, urgency is also experienced elsewhere. In fact, both in the clinical literature (e.g. Karp and Hallett 1996) and as argued at various instances in this study (see Chapter 6), individuals describe feeling urge sensations *outside the body*. Joseph Bliss (1980: 1347) explains this as "a mental projection of sensory impressions to other persons and to inanimate or even non-existent objects". For example, he would perceive a "firm cord running down the centre line of the sheet [of paper]. A need appears to apply pressure to this phantom cord by pulling" (ibid.). He distinguishes between *feeling* an object, not *touching* it:

> At times there is a recurring need while writing to press the pencil point hard against the surface of the paper. A 'feel' is perceived at the end of

the pencil; in my mind, the point becomes an extension of the body, and the 'feel' at the point is translated into a TS-sensitised body site that demands even greater pressure until the point is broken.

(Bliss, Sensory experiences of Gilles de la Tourette syndrome, 1980. p. 1347)

Lance Turtle asserts that this phenomenon, conceptualised as exosomesthe- sia (Dieguez and Blanke 2011), closely resembles feeling sensations through phantoms limbs (Turtle and Robertson 2008). The empty space where the limb used to be is not only recognised to belong to the body, but, for instance in the case of a missing foot, the pressure of the floor against the phantom skin can be felt. In effect, this experience captures a dispersal of sensibilities away from the body, in which sensing capacities extend extracorporeally to take hold in elements of the environment. The formation of sensory con- nections between body parts and the extracorporeal elements that become involved in compulsions then creates a web of static charges between body parts and objects that corresponds with changing spatiality between body parts and elements in the surroundings.

The urgency felt to compulsively touch something requires a particular sensation of a particular intensity that would constitute the dissolving of the pressure behind the compulsive interaction and disrupt the accelera- tion. Whilst Mina remembers what action is required of her to stop the tensions of acceleration when a particular increasing sensation emerges in her throat, it is not enough to diminish the requirement of having to perform it:

Then I wanted to touch it up to after the point of having to throw up (...) that it just about doesn't happen (...) but not ... touching it softly (...) these don't count.

The adequacy of bodily action to offset the acceleration that has to be felt only becomes apparent with the particular sensation that marks the end of the act performance. As such, it does not have an a priori implication to what precise interaction will end the compulsive interaction. This lack of performance direction of the urge sensation was voiced by Ginny:

Ginny: See this jar? Well, I see it's not correctly positioned, and if I place it in the centre – my motivation is that if I place it in the centre I will get a pleasant sensation, and that's why I position it correctly (...) but it's also means that I get the itch ... but ... the motivation ... You see, if I get the itch, that doesn't mean I need to straighten it, you understand?

DB: Yes, yes

Ginny: By way of speaking, it's the reason that I have to do it, but it's also because it feels pleasant, when I do it.

Ginny clearly separates the kind of compulsive performance from the 'motivating' feeling, which implies that whilst the acceleration includes the body and the object, the urge sensation does not entail information on how to perform the compulsion to disrupt the acceleration. In addition to not having a say over the object with which an individual needs to perform a compulsion, this person also cannot determine how a compulsion should be performed. Within the process of becoming, and then being compulsive, human agency is thus severely limited, and the non-human extracorporeal fills this gap. Acceleration is then a subversion of the vitality of the environment as registered in the body, and articulated in movement. Compulsivity can therefore be thought of as an assembled, dispersed phenomenon. Indeed, the urge becomes the sensed articulation of the person being drawn into shared, extracorporeal, non-human becoming-compulsive of the *situation*, which reflects the distributed acceleration of the necessity of the body and its surroundings needing to come together in a *compulsive movement*.

A dispersed acceleration or becoming-compulsive of the situation and the 'ontological location' of the urge can be further interrogated in a series of compulsive interactions that are performed over the course of Siôn's eye-tracking sessions. He compulsively re-places empty and crumbled milk and juice cartons six times before they are permanently binned. The cartons can be seen to be

> *placed on top of the fridge after having been emptied;*
> *placed on the kitchen tabletop;*
> *thrown in the kitchen bin;*
> *taken out and placed back on the table;*
> *placed on a bookcase in the living room;*
> *be put on a small table outside;*
> *and then binned in the large container in the garage.*

These interactions take place in between various other other-than-compulsive acts and happen over the course of half an hour. Whilst they are different compulsions, they can also be considered to produce a series which indicates that the previous act did not dissolve the tension enough, therefore requiring another act. Rather than compulsive interactions accomplishing the cartons reaching the new place – i.e. a result of the compulsive interaction – these interactions accomplish the removal of the cartons from their current situation – i.e. an evocation for the compulsive act. In other words, the urge to compulsively re-place the cartons incited an interaction to change the current situation without immediate reference to any future situation other than a difference *as such*. Nonetheless, the urge seemed to have dissolved enough with Siôn re-placing the cartons, which made the situation less compulsive until it had changed to such an extent that it accelerated towards compulsion again. The urge concomitantly emerging again with the cartons in the new place compelled Siôn to create a new situation again.

Some situations that become compulsive have an immediacy to them that makes the requirement to attend to the compulsive interaction resemble a realisation of what is happening in that moment. For instance, Sara's first eye-tracking video shows her turning around the cleaning cloth in her hand because "it didn't feel right in [her] hand"; when she held the cloth, all attention was called towards this unpleasant feeling, which was then immediately rectified. Sage explains this immediacy as an intense awareness of the fabric of fresh laundry she puts on a washing rack during her first eye-tracking session:

> It's a kind of conscious feeling what you're feeling in that moment, like with meditation exercises, you know. This, that sensation, that conscious feeling, that's what I think I do a lot in the moment that I hold an object with my right fingertips. (...) It can be that I was very conscious of those other clothes pegs, but that I just didn't emphasize it, so it can be that in your eyes I just re-place an object, whilst in that moment I definitely have that focus on that feeling.

As this 'focus on that feeling' interrupts other sensations, in the explosive acceleration, the urge is felt as an acute awareness of the materiality, spatiality, and embodiment of the situation. This explosive acceleration also shows with clean clothes that are transferred from the laundry basket to the laundry rack to dry; when wrapping around the curves of Sage's fingers after she picks one up, the path of the cold, damp, soft flowing fabric of several socks to the drying rack line is immediately subverted to her upper lip, tongue, and underside of her nose. After the shock of the touch with the compulsive moment being most powerful, and the warmth, form, and dryness of the part of Sage's face that touches the sock overflowing in the sock, and the cold damp fabric of the sock overflowing into her face, the urge sensation dissipates and the sock is moved to the drying rack line. The presence of the urge sensation thus differs between compulsions, with clinical research suggesting that relatively long and complex compulsions tend to be preceded by them most often, compared to simple tics (Eapen and Robertson 2015, McGuire et al. 2015).

Whilst other participants in the study often feel urge sensations, for Joe, the situation accelerating into compulsion is never accompanied by a build-up of discomforting tension:

> I get the tic or feel compelled to do something – then I do it. But I don't really feel it arriving. It's just ... 'You experience it: you feel compelled to touch it: you do it'.

Such rapid accelerations into compulsive interactions become possible with touching objects for other-than-compulsive reasons that concomitantly 'escalated'. This happened across compulsive situations and with all participants, for instance when Sage was cleaning small items during an eye-tracking session. After chatting with me, whilst holding a small statue of a Greek goddess, she raises it to her upper lip and nose a few times and pressed it with specific pressure against her skin. More objects followed suit before the end of the session. These items were held slightly longer than the items that did not result in compulsive engagement. With expanding tactile knowledge of the object whilst holding it, or attuning to its synaesthetic textures more finely, the situation becomes compulsive, granting her very little time to halt the acceleration. The body-environment landscape is thus not 'just' tense and full of potential, but is ordered through some sort of magnetism that is not entirely knowable to the human.

(Not-)just-rightness

As compulsivity draws on the spatial situation and retrieves some sort of order from a magnetism between the body and its surroundings, from what and towards what does acceleration turn? In other words, how can we further specify what happens in the state of turbulence. In addition to tingling sensations, the urgent experience of acceleration towards compulsive action can be described feeling an existential 'incompleteness', as a participant in Miguel et al. (2000) elaborates:

> Worse than the obsessions is the feeling that there is always something missing of myself. Very rarely I get rid of this awful feeling that I am not complete, that I need to do something in order to fulfil myself.
> (Miguel et al., Sensory phenomena in obsessive-compulsive disorder and Tourette's disorder, 2000, p. 153)

This incompleteness is more generally capturable by a sense of 'not-just-rightness' that the compulsion brings to a sense of just-rightness. There is no allusion to a moral or even a reasoned rightness, but a sense of rightness that remains unqualified (see also Ferrão et al. 2012). Solving not-just-right feelings requires following an incomprehensible "different kind of logic" according to Sage. Ginny affirms: "I do have my own kind of logical balance, but I don't really understand it *laughs*". Sage makes a distinction between an "OCD norm" which according to her requires "everything to lay in straight lines or on colour", and another one, which "is more a kind of feeling" that alludes to a rightness.

Challenging invocations of this rightness as having a universality, this rightness is expressed in a plethora of ways, for instance through a rightness in some sort of order of a body and a space. In her kitchen, Sage keeps her body 'in balance' with the space; when she turns between the dishwasher and the sink and makes a 360-degree turn, she freezes mid-reach as she realises that this disturbs the balance. The situation thus induces a not-just-rightness. This incites her to make a 360-degreee rotation in the other direction before she can pick up the next object from the dishwasher. A compulsive body-object interaction that corrected between a not-just-rightness and a just-rightness also emerged between a pebble and Elisa's shoe when she accidentally stepped on it whilst standing in front of her house. She had to reposition her foot and push it back into the pebble to erase the unequal not-just-rightness and make it just-right. Additionally, compulsive corrections between objects can be evidenced by Bill grocery shopping. There are just-right locations for new items in his shopping basket, all pinched and pushed into the right constellation: it resulted in a punnet of vegetables to the right, a packet of shrimps on top of it, a juice carton next to the vegetable punnet, and a bag of prawn crackers pushed below the vegetables. A sound-based compulsion that had to be just-right involved Rhys changing the volume of music on his computer. It needs to take place in the right way, as he contends: "if I want one louder, I do two louder and one softer". Also, in olfactory just-rightness, books need to have a right scent, otherwise Elisa thinks:

> Oh no, that one doesn't smell good. I'm not buying that one (…) I'll wait for another version, or I eh … I'll do the pocket version via [webshop name].

Extending her capacity to denote (not-)just-rightness, Sara incites interactions to help her with decision-making, such as when buying a bracelet:

> Then I think like 'which one of the two shall I take, and which one would be the best quality?' Then I'll take it into my hand, and sometimes my hand gets warm, and then I'll think like 'oh' and I'll have that one. (…) I have that with many things, so much choice, but if there's one that's broken, I'll feel it. When I think 'this feels good', or my hand gets a bit warm, then I'll take that one (…) I like bracelets a lot, and then I first feel how it feels, or something, and then I think 'yeah, I actually kinds like that one indeed'.

In these situations, some sort of value is accrued from just-rightness beyond the immediate interaction, which, by extension, reveals some sort of economy in compulsivity.

Noting an acceleration, the period between feeling a not-just-rightness and correcting it to a just-rightness thus forms an experiential indicator of the least turbulence. Whilst in experience and in clinical research, just-rightness is regarded as an *end* to a compulsion (see Leckman et al. 1993, O'Connor 2002, Capriotti et al. 2013, Neal and Cavanna 2013), when it is regarded as signalling the way to halt or disturb the acceleration, it becomes a *mediation*. Considering (not-)just-rightness as mediation draws attention to the collective experience of what Ginny terms situations being "more or less not right", and what Dylan describes as him "constantly searching on what level [he] needs symmetry". Also Sage elaborates that this is a never-ending negotiation, when she remarks on the changing location of a small owl statue on the windowsill:

> It can be that on one occasion, it does feel good when it's on the right, and in another time, the other thing needs to be on the right side.

Even when compulsive interactions involve similar or the same objects, Lowri remarks that creates a world that is "very chaotic (...) indeed always very changeable", and thus consists of multiple changing dottings of varying levels of turbulence. This implies that in compulsivity, the bodily surroundings will entail single or multiple points of emerging not-just-rightness and just realised just-rightness. Following Ginny's description of finding compulsions as enjoyable in certain instances earlier in this chapter, personal geographies of experienced acceleration through (not-)just-rightness then produce a violent mixture of not-just-right 'black holes' that suck energy and attention and just-right 'vitalities' that are pleasurable and energising. In addition to, and in compulsive moments overpowering the various formations of pushes and pulls, fluid formations of desire, compositions of more or less intense (not-)just-rightness then constitute a spatial ordering of subjectivity.

Then, what does it mean when we take away the ontological primacy from the human in compulsivity and refrain from seeing it as contained entirely in the realm of the human: the self, cognition, and the body? What does it look like when the ontological primacy shifts to the non-human, uncouple compulsivity from being a solely cognitive, corporeal affaire and situate it in the affective extracorporeal?

Energy

In making sense of becoming-compulsive and performing unwanted interactions as ways to reduce not-just-rightness and potentially enjoy just-rightness, several participants employed 'energy' as that what is being moved in compulsions. Ginny, Sage, Tomos, and Dylan describe their compulsive interactions as solving 'energy needing to leave the body', otherwise this energy remains as pressure in the body. Sage speaks of "a pleasurable release of something that builds up". Ginny elaborates:

> Ginny: That itch needs to be satisfied, it is a kind of orgasm in a way; then you feel that itch very strongly, for example, and then you crave that orgasm. That belongs together.
>
> DB: And there is also, say, a point of no-return?
>
> Ginny: Yes! That! That's also the case with Tourette's: you constantly deal with a kind of energy that has to do with a kind of orgasmic energy. *laughs* And that is just a very powerful energy that's there, and you can canalise it, or, ehm, that that current becomes weaker, or say like 'well, we're not going to that point', but that energy does have a direction. It wants to go there.

Hence, Ginny understands the involvement of her body in compulsive interactions as supportive of the through-flow of an energy current. The direction of the energy current can be altered to a certain extent, but Ginny cannot block it. In fact, blocking it is regarded "counter-natural" by Sage. When the energy would be stopped flowing through a body, because the body does not cooperate, it becomes stuck. Such a situation results in bodies feeling restless, as Dylan sets out:

> Then you notice that you do have tics, but that it doesn't come out. (...) Often the tics don't reach my legs, to call it that, and then it just stays, indeed, then it keeps bubbling a bit and can't find a way out.

Dylan's body has to perform particular interactions to release the tension, but somehow the body cannot quite find a movement that allows for the energy to 'find a way out'. This happens more often to him, and as an ultimate attempt to shed the discomfort, Dylan immobilises body parts that harbour the pressure by putting them in an awkward position to let them go numb. For instance, he bends his arm around his back and sits against it for a while. It only helps him until it becomes too painful, or when he has to get up.

Compulsivity as mode of energy that is sensed in the body has been considered in the past as well. Combining psychoanalysis and patient descriptions of their compulsive experiences, psychotherapists like Mahler and Luke argued that "symptoms serve the purpose of discharging dammed-up instinctual impulses in a pathologic way. Thus, the tics represent a kind of morbid release, a safety valve for release of tension" (1946: 42–43). Also representing Ferenczi's Freudian theories (1921), this instinctive energy is of an expressive, libidinal kind that, produced by trauma and sexual repression, by virtue of 'suppression' through a denial of bodily movement, comes out as compulsion. Such energic system in these theories is only employed on a sexual 'plane of consistency' (after Deleuze and Guattari 2004 [1980]) in which the bodily surroundings that become involved in these compulsions are reduced to representations of eroticism.

Neurobiological theories do not mention energy as they have a fundamentally materialist nature, but they do employ a reproductive system in considering compulsive acts to emerge from excessive neurotransmitter activity that is incited by the body and 'outside world' and has entered the brain through perception (Leckman et al. 2006). This activity excites individuals and requires them to 'suppress' their movements. The formulation of 'the problem' of Tourette syndrome is a lack of 'inhibition' (Abbruzzese and Berardelli 2003, Belluscio et al. 2011), which in turn suggests a deficiency of surplus activity (Beste and Münchau 2018). Medication that is prescribed to reduce compulsions mainly involves interfering with the reproductive system by disrupting the exacerbating neurotransmitter processes. Cognitive and behavioural studies also denote how individuals subscribe to what O'Connor (2002: 1134) terms a "hydraulic model of tic management", and he exemplifies with "If I keep it in, it will build up and I'll have to let it out later one way or another since I can't contain it". From the latter approaches, behavioural therapies emerged that require 'suppression' of responding to urge sensations and not-just-right feelings. Particularly Exposure and Response Prevention was mentioned to have problematic effects by Dylan, Sage, and Ginny, as they see it as another way of blocking energy:

> And that's fantastic in behavioural therapy[2], it's just that you get stuck with all that tension. Yes, and that's what I told my therapist, like 'we're not doing that therapy', I couldn't take it, I noticed that I had to try and suppress all my tics, and it didn't diminish, it actually got worse, or at least all other issues got worse.
>
> (Dylan)

> Ok, there's something that needs to leave your body and you are blocking it ... and that sounded to me, you know ... that's like with sneezing and stuff ... those are things that just need to get out.
>
> (Sage)

Even when a body would be trained to endure the discomfort of the becoming-compulsive of the body, for instance through behavioural therapy, and develop a resistance to become involved in the compulsive interaction, Ginny argued that the energy reduced to a "small ball" "is really only temporary", and that other-than-compulsive life would re-enlarge it:

> And then that's the problem – that you don't know what to do with the big ball, because you learned to makes it small (...) but the difficulty for everyone is when that big ball is there again.

Ginny explaining how energy is always present, either in a violent 'big-ball' way or as felt in a reduced version, means that the individual continuously has to mediate the situation. Reminding ourselves of Sara's experiences with the small stuffed animals suggests that no matter how often, she needs to roll the seams of the stuffed animals between her index finger and thumb. There remains a tension between her body and these objects and an entirely other-than-compulsive state would then not exist. Indeed, Sage and others suggest that a partial satisfaction of the urge sensations is acceptable as well:

> Sage: If I'm cooking, I know that these hands will touch the pan, so I do it beforehand for a bit// Oh! //If I light the gas, I go to touch the edges like so, ehm, and it's not enough because when it's on the stove they're hot and I still want to, but then at least I got to satisfy it a little bit without getting hurt

> DB: Ah ok ... because you'd still have a recent feeling in your fingers?

> Sage: Yes, yes, and I have done it at least, just ... I still want to, and especially because it's hot, but at least I got to satisfy it a little bit. That's better than not touching at all.

For Sage, performing an as yet other-than-compulsive compulsive act helps retaining a less uncomfortable urge sensation when it would intensify with the heating of the pan.

Rather than as an incapacity, a personal failure to 'control their body', or a fault in the body's functioning, Ginny, Sage, and Dylan see their urge sensations that signal them to 'release' and 'channel' energy as a body purpose and as inherently part of the way in which they relate to the extracorporeal world. This consideration then positions energy as that what is accelerated. Whilst they employed energy as a way to cast some sort of rationalisation over the relations between their body and surroundings and make sense of their sensations and feelings, it is also a fruitful way forward in the development of a theory of compulsivity that is rooted in ontological foundations of fluidity. As a *supporting*, rather than *controlling* involvement of the human resonated with many participants, this situates the body as 'conductor' of energy to come in, flow through, and come out again. How and where energy is detected to pass through the body is then a function of the perceptual processes and body's position in relation to the landscapes of intensifying and de-intensifying black holes and vitalities of (not-)just-rightness[3]. Compulsivity could therefore be regarded as a processual and evental system of reproduction.

This ontology can be developed further by invoking the desirous machinic system by Deleuze and Guattari (2004 [1980]). Given the body and the environment are constitutive and work to reproduce energy as releasing it from the body, the energy is produced by and thus reproduces the production systems. This machinic system relates phenomena through their linkage to the assembled reproduction of desirous energy, the body can be understood to be a site of conjuncture of extracorporeal affects. As I elaborate elsewhere (Beljaars 2019) like other organs, the brain could be considered to reproduce this energy, with specific bodily interaction as the outlet. This would closely resemble the digestive system, which is not considered a normative deviancy, whilst compulsivity is. If the right mode of interaction is not found, or actively 'suppressed', this energy builds up in the body as friction, which is felt in the body as the urge. Indeed, with the situation becoming-compulsive, the energy could be conceived of to locate in the sensory tissues capable of interacting with and extracorporeal entity, and release upon interaction. A compulsion as reproductive event then orders the body and surroundings in accordance with each other and other circumstances. How such reproductions work through various kinds of differences come to matter in compulsive interactions is developed in Chapters 5–7.

Notes

1 She talks about the rubber brand logo glued onto coats just below the shoulder.
2 He refers to Exposure and Response Prevention, which is based on learning to endure the discomfort of the urge, that, with time will diminish in intensity.

3 This invocation bears strong resemblance to the Chinese geomantic practice of Feng Shui. In Feng Shui flows of energy between people and their environments are considered invisible, hence what I argue is that compulsivity is sensibility to this energy system and that people with Tourette syndrome are better positioned to articulate some dimensions of this in with their experiences.

References

Abbruzzese, G. and Berardelli, A. 2003. Sensorimotor integration in movement disorders. *Movement Disorders* 18, pp. 231–240.

Beljaars, D. 2019. Desiring-spaces: compulsive citizen–state configurations. In: Drozynski, C. and Beljaars, D. eds. *Civic Spaces and Desire*, London: Routledge. pp. 189–202.

Belluscio, B.A., Jin, L., Watters, V., Lee, T.H. and Hallett, M. 2011. Sensory sensitivity to external stimuli in Tourette syndrome patients. *Movement Disorders* 26, pp. 2538–2543.

Beste, C. and Münchau, A. 2018. Tics and Tourette syndrome – surplus of actions rather than disorder? *Movement Disorders: Official Journal of the Movement Disorder Society* 33(2), pp. 238–242.

Bissell, D. 2009. Obdurate pains, transient intensities: affect and the chronically pained body. *Environment and Planning A* 41(4), pp. 911–928.

Bliss, J. 1980. Sensory experiences of Gilles de la Tourette syndrome. *Archives of General Psychiatry* 37, pp. 1343–1347.

Capriotti, M.R. et al. 2013. Environmental factors as potential determinants of premonitory urge severity in youth with Tourette syndrome. *Journal of Obsessive-Compulsive and Related Disorders* 2, pp. 37–42.

Cavanna, A.E. et al. 2017. Neurobiology of the premonitory urge in Tourette's syndrome: Pathophysiology and treatment implications. *Journal of Neuropsychiatry and Clinical Neurosciences* 29 (2), pp. 95–104.

Cavanna, A.E. and Nani, A. 2013. Tourette syndrome and consciousness of action. *Tremor and Other Hyperkinetic Movements* 3, pp. 1–8.

Cohen, A.J. and Leckman, J.F. 1992. Sensory phenomena associated with Gilles de la Tourette's syndrome. *Journal of Clinical Psychiatry* 53, pp. 319–323.

Cohen, S., Leckman, J.F. and Bloch, M.H. 2013. Clinical assessment of Tourette syndrome and tic disorders. *Neuroscience and Biobehavioral Reviews* 376, pp. 997–1007.

Conelea, C.A. and Woods, D.W. 2008. The influence of contextual factors on tic expression in Tourette syndrome: a review. *Journal of Psychosomatic Research* 65, pp. 487–496.

Conelea, C.A., Woods, D.W. and Zinner, S.H. 2011. Exploring the impact of chronic tic disorders on youth: Results from the Tourette syndrome impact survey. *Child Psychiatry and Human Development* 42, pp. 219–242.

Cox, J.H., Seri, S. and Cavanna, A.E. 2018. Sensory aspects of Tourette syndrome. *Neuroscience and Biobehavioral Reviews* 88, pp. 170–176.

Crossley, E. and Cavanna, A.E. 2013. Sensory phenomena: clinical correlates and impact on quality of life in adult patients with Tourette syndrome. *Psychiatry Research* 209(3), pp. 705–710.

Deleuze, G. 1991 [1988]. *Bergsonism*. New York, NY: Zone Books.

Deleuze, G. and Guattari, F. 2004 [1980]. *A Thousand Plateaus. Capitalism and Schizophrenia*. London: Continuum.

Dieguez, S. and Blanke, O. 2011. Altered States of bodily consciousness. In: Cardena, E. and Winkelman, M. eds. *Altering Consciousness*. Santa Barbara, CA: ABC-CLIO.

Eapen, V. and Robertson, M.M. 2015. Are there distinct subtypes in Tourette syndrome? Pure-Tourette syndrome versus Tourette syndrome-plus, and simple versus complex tics. *Neuropsychiatric Disease and Treatment* 11, pp. 1431–1436.

Ferenczi, S. (1921). Psycho-analytical observations on tic. *International Journal of Psychoanalysis* 2, 1–30.

Ferrão, Y.A., Shavitt, R.G. and Prado, H. 2012. Sensory phenomena associated with repetitive behaviors in obsessive-compulsive disorder: an exploratory study of 1001 patients. *Psychiatry Research* 19(7), pp. 253–258.

Gerland, G. 2003. *A Real Person: Life on the Outside*. London: Souvenir Press.

Hollenbeck, P.J. 2001. Insight and hindsight into Tourette syndrome. *Advanced Neurology*. 85, pp. 363–367.

Hollenbeck, P.J. 2003. A jangling journey: Life with Tourette syndrome. *Cerebrum* 5(3), pp. 47–61.

Houghton, D.C., Capriotti, M.R., Conelea, C.A. and Woods, D.W. 2014. Sensory phenomena in Tourette syndrome: Their role in symptom formation and treatment. *Current Developmental Disorders Reports* 1(4), pp. 245–251.

Kane, M.J. 1994. Premonitory urges as 'attentional tics' in Tourette's syndrome. *Journal of the American Academy of Child and Adolescent Psychiatry* 33, pp. 805–808.

Karp, B.I. and Hallett, M. 1996. Extracorporeal 'phantom' tics in Tourette's syndrome. *Neurology* 46, pp. 38–40.

Kurlan, R., Lichter, D. and Hewitt, D. 1989. Sensory tics in Tourette's syndrome. *Neurology* 39, pp. 731–734

Kwak, C., Dat Vuong, K. and Jankovic, J. 2003. Premonitory sensory phenomenon in Tourette's syndrome. *Movement Disorders* 18, pp. 1530–1533.

Leckman, J.F., Bloch, M.H., Scahill, L. and King, R.A. 2006. Tourette syndrome: the self under siege. *Journal of Child Neurology* 21(8), pp. 642–649.

Leckman, J.F., Walker, D.A. and Cohen, D.J. 1993. Premonitory urges in Tourette's syndrome. *Journal of American Psychiatry* 150, pp. 98–102.

Lingis, A. 1995. "The world as a whole". In: Babich, B.E. ed. *From Phenomenology to Thought, Errancy, and Desire: Essays in Honor of William J. Richardson, S.J.*, Dordrecht: Springer Science+Business Media, pp. 601–616.

Mahler, M.S. and Luke, J.A. 1946. Outcome of the tic syndrome. *Journal of Nervous and Mental Diseases* 103, pp. 433–445.

McGuire, J.F. et al. 2015. Bothersome tics in patients with chronic tic disorders: characteristics and individualized treatment response to behavior therapy. *Behaviour Research and Therapy* 70, 56e63.

Miguel, E. et al. 2000. Sensory phenomena in obsessive-compulsive disorder and Tourette's disorder. *Journal of Clinical Psychiatry* 61, pp. 150–156.

Neal, M. and Cavanna, A.E. 2013. Not just right experiences in patients with Tourette syndrome: Complex motor tics or compulsions? *Psychiatry Research* 210, pp. 559–563.

O'Connor, K.P. 2002. A cognitive-behavioral/psychophysiological model of tic disorders. *Behaviour Research and Therapy* 40, pp. 1113–1142.

Petruo, V. et al. 2018. Altered perception-action binding modulates inhibitory control in Gilles de la Tourette syndrome. *Journal of Child Psychology and Psychiatry* 90(9), pp. 953–962.

Schunke, O. et al. 2016. Quantitative sensory testing in adults with Tourette syndrome. *Parkinsonism & Related Disorders* 24, pp. 132–136.

Sutherland Owens, A.N., Miguel, E.C., and Swerdlow, N.R. 2011. Sensory gating scales and premonitory urges in Tourette syndrome. *The Scientific World Journal* 11, ID 986538.

Turtle, L. and Robertson, M.M. 2008. Tics, twitches, tales: The experiences of Gilles de la Tourette's syndrome. *American Journal of Orthopsychiatry* 78(4), pp. 449–455.

Van Bloss, N. 2006. *Busy Body. My life with Tourette syndrome*. London: Fusion Press.

Wang, Z. et al. 2011. The neural circuits that generate tics in Tourette syndrome. *The American Journal of Psychiatry* 168, pp. 1326–1337.

Williams, D.W. 2005. *Autism: An Inside-Out Approach: An Innovative Look at the Mechanics of Autism and its Developmental Cousins*. London and Philadelphia: Jessica Kingsley Publishers.

Woods, D.W. et al. 2005. The premonitory urge for tics scale PUTS: Initial psychometric results and examination of the premonitory urge phenomenon in youths with tic disorders. *Journal of Developmental and Behavioral Pediatrics* 26, pp. 397–403.

5 Configurations

Compulsive bodies

In compulsive moments, something is lost. Despite certain possibilities for the prediction of interactions having to be repeated; bodily action is often embarked on in the absence of knowledge about the end result. And compulsions *will* happen, whether accelerations mark a dissipation of rational life and attentive modes are called to the here and now or not. Overlapping dimensions of compulsivity involve correction of a not-just-rightness and/ or an achievement of a just-rightness, a need to rid a sensory urge, and a need to find the best way to channel energy coursing through the body. The invocation of energy provides an understanding of the experience of becoming-compulsive for singular compulsions that sheds new light onto subjectification processes. In particular the spatiality of the relations between the corporeal and the extracorporeal that challenge "radical discontinuity" between the human and non-human (Murphy 1995: 689); a position that underpins neuropsychiatric and psychological scientific analyses of compulsivity.

If present, urges and not-just-rightness feelings are the strongest experiential indicator of the acceleration taking off and compulsive interaction imminently unfolding. Indeed, the sensation underlying them announces the complicity of the body and the extracorporeal elements that become constitutive of the compulsive interaction. The body is not a driving force but performs as a *catalyst* of a situation accelerating into compulsion. This chapter traces the ways in which the body comes to appear in compulsive interactions, how it becomes mobilised through its sensibilities and materialities, and how it is swept up in its vital surroundings. Indeed, the performative approach taken here may suggest an unrelentingly active and guiding role for the body in compulsions, but his chapter demonstrates that this is only a part of the study and that the situation is more complex.

Further unpacking the role of the corporeal in the interchanging other-than-compulsive and compulsive situations, the body seems to follow the reflexive processes in denoting a kind of *withdrawal* from the other-than-compulsive situation. Arguably, this withdrawal analytically captures the experiential state of being made to perform compulsions by another subjectivity, in particular the slave driver mentioned in Chapter 1. It is in this bodily

DOI: 10.4324/9781003109921-6

shift and experiential withdrawal that chemical and sensory and movement management interventions to prevent it grab hold. Indeed, by tracing the shifts is becomes clearer how these different interventions in and on the body work and start to explain what makes them more or less successful in their prevention task. Tracing how these shifts and the bodily withdrawal are intervened with also unveils how individual compulsions have legacies. Attending to the shifts and withdrawal conjures up broader questions about the relations between the corporeal and the extracorporeal and about the ontological position of the human in accordance with the non-human. In drawing this further, it responds to the question; what does involvement of the body in compulsive interactions disclose about immediate and enduring corporeal relations with the extracorporeal?

Shifts

The body's involvement in the acceleration towards participating in compulsive interactions seems to resemble a hijacking of the body parts that are part of the interaction. In addition, it is expressed through a hijacking of people's attention in ways in which people are conscious of as well as in ways that seem to leave people spellbound. In the acceleration towards a compulsive interaction, the body as a whole figures in several other ways as well. One marks a contestation between the fleshy boundaries and the placement of these boundaries as suggested by the sensory realm. Sage recalled not feeling comfortable in her jeans or trousers and that she kept pulling on them to make the uncomfortable sensation go away. However:

> Then I was in the sauna *laughs* and I wasn't wearing anything, and I still had that feeling, and then I thought 'ok, so it isn't the clothes' (...) that every time I just had that feeling like 'there's something on it', but the other day I wasn't wearing anything, and then I thought 'hey, my skin just doesn't fit comfortably!'

Whilst the tactile sensory registers of the body are assumed to cohere with, or at least not be at odds with the perception of the relations between the body and its surroundings, Sage's experience of skin of her leg pulling says otherwise. Besides not quite being able to understand what was needed of her body, it called into question how tactile sensations formed a trustworthy account of her corporeal situation amongst extracorporeal entities. Where do the corporeal and the extracorporeal begin and where do they end? For Sage and others, this marks an ambiguity as to the clarity with which the acceleration can be recognised, or put differently, the body can already become involved in compulsive accelerations without the sensorium registering the transition. Many compulsions described or shown by Lowri demonstrate this. She explains that she very often "gets stuck" at what she was doing with a purpose in mind and coming to the sudden realisation that certain sensations may just have prompted her body to partake in compulsions. For instance, getting dressed in the morning often catches her off guard when composing an outfit slips into the need for her clothes to feel just-right on her body, which can take her up to an hour. The sudden realisation of having 'transitioned' into a compulsive acceleration marks a shift of the terms on which the body is situated, through a sudden diminishment of pre-existing purposes and meaning. Rather, it is a shift towards what John Wylie (2010: 104) describes as a:

> ... world of incessant non-personal, pre-personal and trans-personal relations of becoming, currents of intensity and affectivities – a world which, in its ongoing creative evolution, refuses to ever really settle down into more familiar patterns of subject and object, animate and inanimate, cause and effect.
>
> (Wylie, Non-representational subjects?, 2010, p. 104)

Another way the body figures in the acceleration towards compulsive moments entails a momentary immobility, or as Ginny and Alan describe: some sort of bodily 'freeze'. When Ginny wore the mobile eye-tracker during the study, her body also seemed to 'shift' in the supermarket. She requires all her concentration for finding and picking up the items she had listed, as she methodically paced back and forth and did not look around, barely noticing other people around her. In one instance, crouched in front of the shelves she stares at the boxes of chocolate sprinkles that fill her field of vision. Seemingly completely immersed in the collection of colourful boxes before her, she remains unaware of a woman who wants to pass by her, as Ginny and her cart block the aisle. The woman asks twice before Ginny can withdraw herself from her fixation and jumps up, still looking at the boxes. Immediately after having cleared space for the woman, she crouches down again for a while before taking a box off the shelf. The sight of the chocolate sprinkles boxes on the shelf requires so much of her attention, they completely captivated her, locking her body into the situation.

Experiencing life slowed, body stilled, and being completely immersed in the surroundings is an experience Alan is also familiar with. During our walk on a country path leading through trees and bushes, he remembers having helped scouts find the 7-centimetre-tall red plastic triangle that part of their assignment. Enacting spotting the triangle by pointing at it and walking towards it, his eyes wide open and his gaze fixated on it he remembers:

> It was as if that triangle was completely visible as if it was lit up, and the surroundings faded away. That's how I envisioned it. I always have that, I search hard for something and I can't find it, but when I do find it, it just seems like the sight freezes.

This element had been so captivating that the 'surroundings faded away', and he became completely immersed in the 'frozen sight'. With the excitement of the scouts and them thanking him, the captivation had broken. The triangle in the forest and the boxes of chocolate sprinkles in the supermarket seem to hijack Alan and Ginny's bodies in accordance with these intensities, effectively pausing representational life until something or someone breaks the affective tension and which stops the acceleration. These situated objects trigger the acceleration towards a compulsive situation through inducement of a forceful 'attunement' to the situation that is so immersive that it impedes on bodies' ability to attune to anything else. It is then plausible that attunement continuing and the body remaining 'enlaced'[1] in the situation can be explained by the person involved not understanding how their body should proceed to perform the compulsion. This creates a deadlock-like situation in which the person seems to be captured in the increasingly compulsive situation but does not immediately know how to break out of it.

To further explore this bodily capture in the situation further, we turn to theories of subjectivity. Such theories argue that subjectivity is fundamentally co-constituted by the human (or non-human animal) and its environs in more or less equal measure. Instead of human action being entirely rooted in the psyche and free will, Braidotti (2011) citing Deleuze and Guattari (2004 [1980]) argues that it is situationally composed: the material and immaterial elements (e.g. air, light, temperature) that make up one's *territory* or *milieu* frames and guides, for an important part, what is (im)possible. Given that compulsions are so strongly experienced as lacking human initiative and the psyche does not seem to be involved all that much, tracing the 'capture of the human subject' in compulsions provides more insights into the transition process between compulsive and other-than-compulsive situations.

A parallel can be drawn between the compulsive body and the chronically pained body, as according to Bissell (2009), it shares the experience of capture, felt as immersive or totalitarian. He argues that "chronic pain is presented as an undesirable affective intensity that has no recourse to intentionality and meaning but *territorialises* the body in ways that prevent other intensities from taking hold" (Bissell 2009: 911, my emphasis). Territorialisation is a process of affection across difference that produces and reproduces orders, developed by Deleuze and Guattari (2004 [1987]). Bissell argues that pain is such a powerful bodily order that the body *de*territorialises from further life and can only be responsive to the pain. Medication that reduces the pain is then a way to break the deadlock and *re*territorialise the body to regain purposeful and meaningful life. In turn, understanding compulsive processes to take place through de- and re-territorialisations helps demonstrate that compulsivity is a powerful order that it can pull people away from other-than-compulsive situations and diminish when situations become just-right again.

Territorialisation processes are based on a Deleuzean understanding of Bergsonian theory of connection through 'images' (Deleuze 1991 [1988]). Images of metaphysical bodies[2] produce the potential for affection and a change of order, in which affection occurs through metaphysical bodies 'becoming-other'. As Deleuze and Guattari (2004 [1980]) exemplify, the wasp's and the orchid's reproductive systems require them to interact, for which they need to 'form' themselves in the image of the other. The orchid forms itself in the image of the wasp and becomes-wasp, and vice versa. In the body's territorialisation towards the compulsive moment, it may then be conceived to 'become-other'. For instance, in compulsively pressing one's finger into a tip of a table, the finger would be considered to briefly 'become-table'[3] and the table to 'become-finger'. This does not entirely capture the compulsive experience. Building on this and to capture the specificities of the body's transitional participation into compulsions, the experience of being configured towards a compulsive situation may resemble autistic perception. Autistic writers suggest that they perceive the non-human environment to be incredibly complex and differentiated, which warrants one's full attention:

> Autistic perception dances attention, affirming the interconnectedness of modes of existence, foregrounding the relationality at the heart of perception, emphasizing how experience unfolds through the matrix of qualitative fields of overlap and emphasis already immediately moving toward expression in a dynamic field of becoming alive with co-composition. For autistics, language comes late.
>
> (Manning & Massumi, *Thought in the Act: Passages in the Ecology of Experience*, 2014, p. 6-7)

Rather than an absence of meaning, environments explode with liveliness. The freeze experienced during a compulsive situation may not be capturable in words due to the multiple calls for a response placed on the body. Alan's forest surroundings could not be attended to, because the red triangle required all his capacities, overwhelming him with possibilities. Compulsive territorialisation thus seems to be all encompassingly vital. During a compulsion, the human body thus does not form in the image of an object, rather, a more accurate analysis suggests that it *reconfigures towards a compulsive situation*. Compulsive configurations are then more or less encompassing bodily reorientations that are invariably followed by a reorientation of subjectivity towards a compulsive situation that can only be attended to through performance and feeling, not reasoned knowledge and/or memory.

The recurrence of bodily configurations towards affective intensities becomes apparent in Sage's experiences when she takes a walk in the rural area outside her home. She feels her gaze being drawn towards pointed edges:

> That I don't pay attention to what I actually – where I actually am, you know (...) I miss it sometimes, that I can't just look around me for a bit, because I'm always focussed on those pointed edges.

By having to make a conscious effort to divert her gaze from pointed edges to 'pay attention to where she is', her body seems to be pulled towards the compulsive situation that is accelerated by seeing these edges. With the becoming-compulsive of the body, it seizes to be configured in accordance with other-than-compulsive life. Clearly, it produces difficulties for her to resume to a preferred 'looking around', even if she 'suppresses' the urge to push her finger into these edges. When she walks around in her flat, the same happens; the mobile eye-tracking recordings regularly register her gaze flashing to pointed edges of objects and the indented corners of the room. Especially when she walks from her kitchen to the living room, Sage's gaze meets a few particular points in quick succession. It demonstrates how the surroundings emerge as a patchwork of 'accelerant intensities' that demand particular engagements in accordance with particular corporeal situations. When she cleans the top of the coffee table, hovering over it one of its sharp corners draws her attention a couple of times, which ostensibly starts the compulsive configuration. The situation accelerates, and immediately after finishing wiping the table, she compulsively presses the tip of her right index finger into it. The changes in the bodily constituencies then place new configurational demands on the body.

Thus, in the compulsive configuration something may appear to be lost; a sense of self, some sort of rationality, and sense of purpose. However, thinking through the configuration in such terms does not do justice to the experience, nor is it helpful for the non-normative analysis of compulsivity committed to in this book. Compulsive moments are not inferior to other-than-compulsive moments; consciousness just shifts into a kind of mesmerism, identified by Alan and Ginny amongst others. Indeed, where in compulsive moments reasoned, abstract life disappears and the body and sensations come to the fore, the reverse happens in 'rationalisable' moments from which the body and sensory life disappear (Leder 1990).

In the absence of a sense of self in compulsive interactions, including the regular absence of direction for the performance, intentionality in compulsivity does not emerge fully with the human. That is, except for finding the easiest, quickest, and most comfortable way to configure back to other-than-compulsive life. Instead, intentionality in bodily action is inherently distributed with the compulsive interaction being rooted in the body's conjoining with the vital extracorporeal world. Exploring how the human and non-human are entwined through embodied life, Alphonso Lingis (1998) argues that the world exists in the interrogative mode, and to be human is to respond. Intentionality in compulsive interaction can be understood better following Lingis' theory. Responding is only possible through finding orientation according to the elements that create worlds, and what these elements emerge with this orientation. He conceptualises the extracorporeal affects through 'levels'. In searching imperatives for action, levels denote matter in terms of light, colour, sound, texture, as bright, dull, vivid, refractive, smooth, and scratchy. Such levels of materiality intertwine with sensibilities that allow humans to *see with and according to* the levels of matter. Lingis explains this according to the following: with the sun setting and the levels of light changing, landscapes filled with individual houses and trees increasingly clump together and first transform into new shapes in accordance with their lightest parts and eventually fade into the grey, dark blue, and black night world that demands altered responses. All presences thus fold towards each other; they appear in accordance with each other and form the subjectificational context with which the human corporeal conjoins (after Wylie 2006, Lea 2009; Ash and Simpson 2016, McCormack 2017). The spaces in-between create the potential to communicate (Wylie 2006, 2009, Harrison 2009).

Vital landscapes with which accelerations take place are then an enduring questioning. Accordingly, being drawn into the compulsive configuration could then be seen as feeling the call placed on one's body, and the compulsive interaction, a mode of response. This all helps to determine how some situations start to become compulsive and how acceleration takes off; for Sage, the rural landscape with the pointed edges and her compulsive need for the 'punctuation' that is also mentioned by Bliss (1980). This makes even more clear how compulsions are experienced as meaningless. Them being co-constituted therefore removes the possibility for meaning, rationality, and purpose. Such sense-making efforts can only ever be applied afterwards.

Whilst instances like Sage's unveil a compulsive configurational pull on the body incited primarily visually, other ones involve touch. Anything that touches Elisa's body does not only remain in her awareness, it also holds great potential to incite rapid compulsive configurations that are very difficult to 'break'. Her body easily 'sticks' in such compulsive moments and it takes a lot of effort to configure back towards more intentional life. Therefore, she does not wear underwear nor glasses. She also does not wear piercings as she would play with them until they would tear her flesh. Also, she can only sleep well in 100% real silk bedding; lying in synthetic silk bedding will always be not-just-right and prevent her from relaxing enough to fall asleep. The human body *being touched* without explicitly moving towards the extracorporeal element thus seems to have a similar effect as the human body *moving towards* such objects. As one can often not escape such interactions as often people need to wear at least some clothes, these interactions retain the body in its compulsive configuration in accordance with the extracorporeal materiality that touches it. For Alan, this feels like his body is hijacked and leads him to prefer being naked, especially as in the shower he does not have compulsions and only very few in bed. The spot where Alan's body is touched immediately starts itching:

Alan: If someone unexpectedly rubs my back, that spot will start to itch

DB: and it stays sensitive?

Alan: Yes, for a bit, a few minutes or so I'll be itching

> DB: Can you remedy that?
>
> Alan: I just scratch for a bit.

For Alan, the itching also occurs when his hands are tied in a position that refrains him from scratching. This presents real problems when he walks his dogs, as every now and then, when the itching sensation is too strong, he needs to stop walking and transfer all leashes to one hand to free the other for scratching his cheeks and nose. This itching is another manifestation of the urge sensations and indicates the body configuring rapidly towards a compulsive interaction. This immobilisation of the hands reproduces the immediate vulnerability of the body when they are unavailable for scratching. Similar instances were reflected by other participants, and it transpires that certain bodily situations, especially those that are restraining (both psychically and disciplinarily through not being allowed to do something), may make the body more prone to become compulsively reconfigured. Activities like dog-walking and carrying plates with cutlery to the kitchen characterised by this problem have particular geographies and need to be carefully considered before they are embarked on.

The body's configuration towards compulsions is intimately entwined with the objects that become involved in the interactions, which are perceived and felt in several ways. Building on Deleuze and Guattari's (2004 [1980]) theories of affect, corporeal configuration towards a compulsive situation is reliant on an incessant anticipation of being affected by extracorporeal materiality, in which the extracorporeal is already implied in the corporeal. During the interview Ginny puts this anticipation in words, discussing the honey jar she had just slid out of reach to the far end of the table after having cupped her right hand around it and squeezed it tightly. Struggling to concentrate, its presence was distractive for her, and she kept feeling the urge to squeeze it. She explains, and whilst she speaks, her hands cup around the virtual jar and she tenses her muscles to squeeze it:

> Now you're talking about it, then I immediately get … then I see that jar and then I feel that jar". Many things are sensorial things. I memorise very intensely how things feel sensorially. (…) I get stuck in it, you know, because I can already feel it.

Instead of an extracorporeal presence that itches, this compulsive situation marks the absence of the object. Ginny's hands are anticipating the jar and her body has reconfigured in accordance, eagerly awaiting the actualisation of the sensation of the jar. The anticipation of bodily engagement with the object in compulsive fashion thus reflects an affective intentionality rather than a cognitive intentional tendency (after Levinas 1998, Bissell 2009, Ash and Simpson 2016). Hence, even in the absence of objects with which compulsions have been performed, in the anticipative mode, the body retains the capacity to become configured into virtual compulsive interactions.

In this instance, Ginny needs the jar to be out of reach to prevent actual compulsive interactions from having to be performed, whereas Sage remembers that in her teenage years she needed objects to be nearby. She was frequently unable to sleep without certain objects on the carpet next to her bed. These objects could be "anything that caught [her] eye" that evening, ranging from stuffed animals to coasters to small statues. They had to stay in close proximity when she slept, to refrain her from lying awake feeling not-just-right. Sage's body had been configured towards these proximate items in the situation of going to bed, and the only way to 'break it' and continue other-than-compulsive life would be to present the objects to stop 'pulling on her'. The spatiality of her body in her pyjamas amongst the bed, particular items in the room visible from her pillow, as well as a state of tiredness, bedtime, closed curtains, and the fresh taste of toothpaste in her mouth in addition to an inexhaustible list of elements that produced the situation becoming-compulsive in this particular way. Only with the right collection of objects at the right proximity of her body, the compulsive element of the situation could dissolve and her body reconfigures for sleeping purposes.

In conjunction to so many other compulsive interactions discussed in this book, the necessary constellation for Sage to remain other-than-compulsive neatly reveals the intensely spatial dimension of compulsive sensibilities. Reminding ourselves of the virtual landscape of vortexes or black holes and vitalities of energy that is ordered in accordance with (not-)just-rightness, it is clear that compulsive situations towards which people configure are not perfectly spherical. Rather, their geometry forms in accordance with the boundaries that are dictated by the spatiality of the sensorium; the mapping of sight, (im)possibilities for touching extracorporeal elements, as well as soundscapes. The latter is demonstrated by Mina's compulsive requirement to close the bathroom door in a way that it produces a dry clicking sound that is just-right, before she can continue going (back) to bed or continue her daytime activities. She holds her ear close to the mechanism that produces the sound to make the judgement. The position of the body in relation to elements of its surroundings is thus crucial in the way people can be 'lured' into compulsive interactions. The compulsions mentioned in this section thus unveil how the body in action is pushed further into a radical yet tangible passivity (after Harrison 2009). Indeed, the configurational understanding of compulsion radically reconfigures the human involvement to the compulsive interaction and takes away its ontological primacy.

Detailed knowledge about situations and the flexibility to change situations through various reorganisations, such as the set-up Sage needed to sleep when she was younger, often become mobilised by participants to find the least discomforting and interruptive ways of accommodating anticipated compulsive engagement. Minimising the 'risk' of becoming configured towards a compulsion, and maintaining the optimal state of non-compulsion and/or being pleased by just-right situations then creates a compulsive libidinal economy of well-being. Taking a closer look at how negotiations of configurations and strategies to minimise the risk to create the conditions for compulsive interactions provides new insights into the formation of well-being.

Losing bodies

The potential for, and incessant threat of a reconfiguration of the body to attend to the compulsive situation instead of a chosen one articulates in the ongoing mobilisation and experience of the body outside compulsive situations[4]. For instance, Ginny recalls not being able to configure back to other-than-compulsive life when she leaves the house for work, knowing that objects are not in the right place: "sometimes I get an acute error, and I can't think anymore". Failing to configure back to an other-than-compulsive state before leaving a compulsive situation can result in difficulties aligning the body with tasks at hand. Such experiences lead some to question that they have their body 'at their disposal' in other-than-compulsive situations. Dylan reflects the difficulties he associates with it, in particular, he regards his "fine motor movements" to be compromised:

> It's the fine motor movement that requires lots of calibration and lots of touching. It's inevitable that I touch the tap with my fingertips, with my whole hand. It's just not very handy. Would you like bumping your fingertips into it and then jerkily grab hold of it? That's unnatural.
> *he moves his hand to the tap and turns the handle so the water starts flowing*
> Here too, open in one go. Imagine missing [the handle] when you reach for it in a fluent motion, imagine missing it. You'd have to recalibrate that whole fluent motion ... That's what stresses me out, that stress is just always there when I do fine motor activities ... Water flows over my hands ... it's even a bit painful ... painful is not even the right word, but it feels strange ... something flows over your hand (...) This fine motor bullshit, and then having to wash your hands all the time. How tired that makes you – terrible!

Dylan discussing moving his body in terms of calibration suggests he needs to actively steer his bodily movements, as he cannot always trust his body to move in unthought ways. The physical discomfort of bumping into things is one thing but more so being touched unexpectedly tends to trigger reconfiguration towards compulsive situations. Indeed, stopping such configurations may be especially strenuous when he needs to interact with extracorporeal materiality in other-than-compulsive ways. Objects might become so affective that these interactions require a very deliberate and careful approach. Also, during the eye-tracking sessions, he often had to resume doing the task he had agreed to do for the study. Picking up objects to put them in a new place often resulted in him getting stuck into playing with it and talking about it, until he would reconfigure towards the task and continue with it. Movements contributing to other-than-compulsive action can thus both incite and retrieve a body from compulsive configurations.

Also circumstances in which other people are present can create unprecedented situations. To ensure that no compulsion is performed that may create unwanted social tensions, most people will be more attentive to their sensations and signs of configuration so that these can be incorporated in seemingly wanted movements, hidden from the sight of the onlooker, or endured without performing the compulsion. Indeed, Sage remarked that indeed my presence had such effect when I visited her place for the eye-tracking sessions: "The itch is there the whole time, because that's always the case with you here. I don't want that, so sometimes I suppress it". In these situations, my bodily presence becomes part of Sage's bodily constituencies which she experiences as catalysing her bodily reconfigurations towards compulsive engagements. Quite the opposite happened during my meetings with Elisa. She had planned to "let it all out" and show me all her compulsions during an observation session. However, not everything did "come out" and she was unsure why. Having had to restrict her bodily movements around other people her entire life may have been anticipated in a non-negotiable way. Indeed, also Nora could speak relatively freely about her compulsions, including those she was embarrassed about and about how she negotiates being in groups of people by "letting the compulsions all out" in the bathroom. She also needed to do this during our meetings, only showing a few compulsions: "even so, without noticing you still hold them in". There is much more to the social mediation of compulsivity, which is further elaborated in Chapters 7 and 8.

Losing one's body to the compulsive configuration is not always experienced as thoroughly problematic, and situations can be enjoyed because they are just-right. For instance, Alan feels thrilled when he sees rows of trees aligned perfectly every time he passes them, and will stand still to really look at it. Also, Sion has designed and created the almost entirely white home office to fit precisely with several elements of just-rightness, for instance the levels of light, the alignment of the furniture, and the connections between skirting boards in corners:

> So I can also just very much – when I made this [room] for instance, it's changed completely. The walls are treated, a new floor, the desk I made, a new couch, and when it was finished, I could just stand here, very quietly. I could observe for hours. Really bizarre, that you enjoy that tranquil sight. My eyes, they … it's quite odd of course that … I get it though, I can explain it, but that you can be so intensely happy that your eyes have that tranquillity.

The enjoyment of becoming enlaced into certain compulsive configurations and staying in them for a while does not make any other compulsions less disturbing or less painful, nor was the element of enjoyment shared across the participant group. Nonetheless, being swept up in these situations and being absent in other-than-compulsive life for even a moment can be welcomed.

Activities that cannot permit Dylan's body to reconfigure towards a compulsive situation because it would put him in danger, such as driving, are negotiated through a deliberate focus on the body. Similar to strategies employed by people with autism (see Cowhey 2005, Lawson 2005, Davidson 2010), he uses a special rubber mouthpiece onto which he can bite down during the drive. Biting on the mouthpiece helps with reminding him of his body in the sense that he actively retains it in the configuration necessary for driving:

> I also use it when I have to concentrate. When I am vacuuming, I am wearing it. (...) At the point you can shove tics out in the mouthpiece, the storm in your head stops.

Such interactions might help channel extracorporeal affects in such a way that for a short period the body is insensitive to the pull to shift into compulsive configurations. In the supermarket, he employs a similar strategy: he pushes his shopping cart, walks past it, stands in its path, and lets it bump into his backside, upon which he continues searching for the items on his shopping list. Such other-than-compulsive interactions thus seem to be purposefully incited to reduce the requirement for the body reconfiguring into compulsive situations when the person needs their body to be configured for other-than-compulsive activities. Not using such strategies, nor using the right strategies may cause difficulties with relating to one's body. Ginny seems to have such difficulties. She regularly goes to have a massage to "get back in [her] body", and situate her body as belonging to her, almost 'reconfirming' its materiality and capacities. This seems necessary as configuring her body back to other-than-compulsive movements does not always work; or its compulsive reconfiguring is so relentless that her corporeal existence does not always appear distinctly different from the extracorporeal:

> Ginny: I have to experience matter. I don't experience matter, that's how it is I think. (...) then I just have to touch everything, I'm just all over everything if I let myself do it, yeah.

> DB: What do you mean with not experiencing matter?

> Ginny: Well, if you often – if you experience yourself in unity and as very large, not limited to a body, then you surpass the dimension of matter, then you're in a different kind ... (...) I've always been; I used to be homesick to [an unknown faraway place] which is a bit traumatic, thinking like 'Jesus, what am I even doing here? It's so boring here and everything is so slow, and I have to walk everywhere myself, and I can, in my, with my soul, or with my everything I'm there in a second!' In matter everything is so slow, so matter has always been a kind of... a

negative thing for me, because it was my resistance and it withheld me from going fast, you see (...) For example, when I go to get something I'm already upstairs and then I often bump into the edge of a wall. I get stuck because I am there already, but my body isn't! *laughs*

In Ginny's experience of some situations, her body does not only attend to extracorporeal materiality but ontologically dissolves in it by becoming 'of a different kind'. Becoming other-than-body – and even other-than-matter – overcomes the configuration conundrum in which the body requires constant reconfiguration in accordance with other-than-compulsive life. As such, Ginny negotiates having to perform compulsive interactions with a denial of corporeal existence altogether, which, in turn, produces all kinds of other difficulties, including an inability to differentiate between corporeal and extracorporeal matter, and an inability to configure back into other-than-compulsive life. The next section lays out how negotiations of performing compulsive interactions hinge on managing and negotiating compulsive configurations of bodies.

Interventions in corporeal affection

The process of compulsive reconfiguration can be negotiated. The current medical conceptualisation of and ontological responsibility over compulsivity is located solely in the body, which reflects in the available therapeutic support that becomes accessible upon diagnosis. Negotiation strategies on offer roughly map onto the brain through medication, and onto bodily movements, sensations, and cognition via behavioural therapies. This section traces the ways in which a variety of therapeutic support interferes with the body to affect the performance of compulsions, and what that says about the status of the body in compulsivity.

In early 2021, no medication is available that is specifically designed to diminish compulsions for the Tourette syndrome population. There are, however, multiple pharmacological treatment options that affect aspects of the phenomenon. Therefore, many people have a history of trying different types of drugs, often resulting in them finding one that works for them at that point in time, having to try another kind, resorting to behavioural therapies, or stop external interference altogether. Ginny's experience with antipsychotic medication was initially good:

[Name medication] was very pleasant in the beginning; you think 'the signal comes in and immediately goes in the right direction. It was like a sort of holiday in my head, you know. When I'd wake up I was immediately able to get up and not have all the thoughts, it was so much easier.

Independent of its capability of decreasing urge sensations to perform compulsions, medication seemed to affect the context in which compulsivity figures in people's lives. Elisa describes changing medication and its effects:

Yeah, I'm feeling much better, because [brand previous medication] suppressed my simple tics, but not eh … my sombreness. That actually worsened. It's one of its side effects that you get more depressed (…) It's not designed to decrease tics, but it does make you apathetic and I get sad because of it. With the new medication I don't have that at all, so that's ideal. I'm not yet at my old level where I need to be, but quite far in the right direction. (…) When I feel better, I'll be more bothered by my tics than when I feel sad.

The new medication did not stop Elisa's body from configuring towards compulsive situations as much as the previous tablets had required from her, but feeling less sombre and more upbeat changed how she experienced her participation in the acts. In this case, anti-depressants exacerbate the difference between compulsive and other-than-compulsive life, in which configurations towards compulsive situations are more noticeable. For Dylan and others, medication diminished the quantity of compulsions, but its alteration of the relationship between the body and its surroundings is more often than not experienced as profoundly disturbing (see Chapter 2). Sage describes being prescribed medication with "increased chances of suicidal tendencies" and being told that "[she] already had them anyway, those suicidal tendencies", which left her perplexed, despite "it being closely monitored". Ginny elaborates on the alteration of her lifeworld:

Ginny: It's being muted, those senses, the neurons are just being shut down. And that had the effect that I felt that I was in the room on my own behind glass, and it all happened over there *points*, and I couldn't feel anymore; I could see it, but not feel it. It was really very strange.

DB: Yeah, so sight wasn't enough

Ginny: Yes, it was as if you're just not involved in the world (…) It made me very depressed, I felt very out of touch with the world (…) I thought 'it doesn't even bother me whether I'm dead or alive', and I have children, you know! Then I just had – then I felt entirely disconnected, from the earth or anything.
(…)
If I don't feel any stimulation, I become very unhappy. (…) I normally experience myself everywhere, that I can go everywhere. With the [name medicine] it was like I couldn't have access to certain places, dimensions, in other spots, you know. Now, I can just sit and daydream and fantasize a bit, but I couldn't do that at all anymore. It's weird, hey! And in the morning you're just sat in the material world, that only this exists; only the chair, only the stuff in the room, and nothing else. So that's very boring (…) I thought it was just awful. And it's ferocious stuff, it is. It took me 6 months to slowly phase it out. I had a kind of pain in my brain, your brain really needs to recuperate."

Ginny's body seems to remain unaffected by the extracorporeal materiality that she could have interacted with compulsively, had she not been on this medication. The experience describes pure sensation without feeling. Instead of only taking away the experience of becoming-compulsive, this medication seems to completely obliterate all sensibilities to become affected by *any* element of the body's constituencies. This includes affections that build up more purposeful, meaningful life. It points towards an incapacity to notice the body being drawn into interactions or to a complete lack of compulsive configurations altogether. The indifference renders the medicated body an object among objects. Such bodies are then brought in a state of numbness as materiality without sensibility, deprived of a vitality assumed inherent to humanity. Indeed, the chemical addition to the body reflects the falling away of a perceptive state that seems to pick up on the excesses of forms, colours, textures, and temperatures that compulsions seem to be an ongoing mediation of. Denoting how only a bland functionalist appearance of the world seems to have remained in medicated states reveals the various forms of intrepid liveliness that are not only guiding situated life, but are vital to life itself.

Often taken when medication did not have the desired effect, cannabinoid additions to the body confer similar but profoundly different effects on perceptive processes. Brewed as tea Ginny explains "it's as if there's a veil over or a balloon coming over me, a kind of protective layer that stops those stimuli from coming in so violently". Smoking medicinal cannabis through a water pipe/shisha device Dylan can "instantly breathe again". He recalls having switched from publicly available "coffeeshop⁵ weed" and waking up the next day: "boom, space in my head, I functioned, I slept fantastically, and I could walk again". With the medicinal cannabis Dylan experienced his body to remain configured towards other-than-compulsive life, configure towards compulsive situations less often, or does not get stuck into compulsive situations as vigorously. Cannabinoid additions to the body thus seem to retain the vitality of the extracorporeal world, whilst diminishing the perceived 'punctuation' of the fragmentary appearance of that world.

Maintaining the body-world formation produced by cannabis requires meticulous self-monitoring and management. For Dylan, the dose of the active ingredient needed to be exactly 0.16 gram, and the smoke he inhales is kept in his body for precisely five seconds. The fact that he is in charge over the doses, and that he can determine what works best for him, moves away from any standard expectation of 'daily functioning', and allows for a more nuanced intervention that he can adapt quickly with the flows of daily life. Running out of cannabis means that he cannot walk the day after. Therefore, having to rely on cannabis to function as well as he does poses risks, as he regularly has to find alternatives of coping because "the pharmacist had been too late, got its order wrong or hadn't received the weed yet". Any type of medication is also applied to negotiate certain circumstances that are known for making the body extra sensitive to become enlaced in compulsive configurations. Nora takes medication in anticipation of noisy situations or those prone to evoke social anxieties, such as performances with her choir on stage. As she argues, in these moments she chooses to give up being "her happy self" to feel less susceptible to compulsive demands.

Getting chemical additions to the body 'right' thus requires a particular understanding of the body, its sensations, and its thresholds. Chemical constitutions of the 'acceptably' compulsive body require an understanding of one's own body from the outside, and in mechanical terms; what additions, how much, and under what circumstances to take to have a desired experience of one's lifeworld. Adding too much, too little, in the 'wrong' way, at the wrong time, and in failing to anticipate a particularly challenging future, the body comes to a grinding halt. The necessity to think about one's body in this machinic way purports it as fundamentally dependent and unruly (Philo 2007, Chouinard et al. 2010, Schillmeier 2010, Hall and Wilton 2016). This feeds into ablest models of embodiment following medical traditions that see bodies configured in accordance with normative standards and employ hierarchies of bodies (Duff 2011, Andrews et al. 2012). Extending this critique from critical disability scholars, this notion renders the body in economic terms in which the 'optimal state' is the body medicated in such a way that it is most productive (Gleeson 1999, Hansen and Philo 2007) only being understood as whole when chemically balanced in such a way that it is most productive.

Another addition to the body that alters perceptive bodily processes is through the consumption of alcohol. Mina experienced more problematic than helpful effects of the different medication she tried, and medicinal cannabis could not be arranged for her. She uses alcohol to "dampen the effect" "when it's too much", despite finding "it not very smart, of course", given the addictive effects. She explains her predicament:

> At some point you have an alcohol problem. Then you have to get to rid of that! I've never taken a sip and now it's kind of becoming a proclivity to at least at one point in the day just to be able to relax, but yeah, this is not a solution.

Similar to Nora's habits of taking medication, alcohol can be added to the body to negotiate situations in which the urge sensations are too disturbing and not feel her body at all. Whilst alcohol and cannabinoids can be taken to offset sudden disturbances, antipsychotics and antidepressants need a six-week 'build up' in the body. Nora was the only participant who was able to use medication incidentally.

Behavioural therapies, in particular Habit Reversal Therapy (HRT) and Exposure and Response Prevention (ERP) do have a momentary effect and can be mobilised when the person feels they would benefit from it. In line with growing concerns over the hit-and-miss success rate and the side effects of medication, and the increasing evidence that behavioural therapies diminish tics, they are becoming more popular and have become the first-line treatment for people with Tourette syndrome in Europe and the

United States[6]. These interventions aim to alter the bodily (HRT) and sensory (ERP) responses to compulsions. In case of HRT, body parts involved in the compulsive act are positioned in such a way that they cannot be performed, but which also seems to dislocate acceleration. Dylan "noticed that [he] needed to stop [his] tics, and it didn't get less. In fact, it became worse, or at least, [his] other issues got worse". He adds that the technique did not refrain interactive compulsions from taking place, potentially because there is not one motion that incites the wide variety of interactions, even if hands and arms would be immobilised, balancing, alignment, and ordering compulsions can still be performed.

In the case of ERP, people are trained to dismiss the sensations that urge them to act compulsively. Nonetheless, having to focus on the urge sensation, and refusing to become enlaced seemed to instil an acuteness of bodily presence in participants in similar ways that pain "holds sway" over a person, with "one's whole being [being] forcibly reoriented" (Leder 1990: 73). This reflects Sage's assertion that focussing her attention to the urge sensation makes her "successful in registering where it is located because of the therapy". Therefore, ERP seems to assert a new striation of the body through the urge sensation, paradoxically intensifying the experience of the body slipping away into compulsive reconfigurations. Indeed, whilst ERP does provide people the power to resist following up on the urge, it does not stop compulsive reconfiguration. How well ERP works for people is then experienced to relate to how well and often one "practices" this resistance, according to Sage. Lowri finds this one of the most difficult aspects. 'Practicing' for half an hour refrains her from doing other-than-compulsive tasks, as she feels that every activity she undertakes has a compulsive element to it:

> Especially with that water I had to practice, that is super difficult because it was mainly a habit to drink in that way. That water, that way, that I had to kind of unlearn it, because to drink in a normal way ...? Yeah, drinking in the compulsive manner was just imprinted in my mind.

Such compulsive interactions that are not distinctively different from other-than-compulsive interactions – "random" as Lowri puts it – are not captured by these therapies. Notwithstanding, for some participants who had tried ERP, it did alter the compulsive enlacing *as such*. For instance, alluding to the sensations, Rhys argues that:

> With the pressure increasing, I can better handle it (...) and that's a very comforting feeling. You have more control where I didn't have that before.

This feeling of decreasing powerlessness is reflected by Tomos, who sees such interventions as overflowing into an attitude to life beyond compulsive configurations. Reflecting on a period in his life during which due to the medication he was incredibly tired and "dead-like", he could not leave the house or undertake many activities. He argues that:

> It is then that you have to [do] exposure [therapy]. I think that you have a kind of wealth as human being when you have limitations. When you overcome those limitations, or perhaps not overcome the physical limitations; I will always have tics, but I can deal with it spiritually. It is then that you become such a character. That's what I want to achieve. I'm not there yet, but I really want to be that.

Tomos' understanding of exposure therapy in his life has become a personal affair; he understands that being able to resist compulsions in the moment makes him more of a character, a more solid person. It exemplifies what others have also implied; acquiring more 'control' over one's compulsive interactions by doing less of them or none at all would make them more fully human. And whilst socio-politically these people do not lose any aspects of their humanity under any circumstances, despite being made to feel that way by onlookers and societal institutions throughout their lives, the experience of individual compulsions does hold truth through the configurational emergence and disappearance in which their bodies collapse into the more-than-human situation.

Non-human environments reverberate through the bodies with such violence that they can lose all boundaries, and people feel they no longer fully inhabit their own body. The alienation of the body that is experienced during the configuration towards the compulsive situation simultaneously reveals a world that is livelier than it appears along reasoned and memorialised lines, and reflects a mere 'witnessing role' for the human (Dewsbury 2003). The intensity of this world paradoxically only emerges with it falling away, stripped bare to its mere functional existence. Compulsivity thus works through a binding of the body with its surroundings as ordered by perceptive processes. Configuration is then a formation or order of affects; it suggests that in compulsivity, affects are not everywhere, they self-organise between the body and the extracorporeal. The next chapter considers how objects and spaces as extracorporeal worlding emerge within compulsivity.

Notes

1 I prefer using 'enlacement' to similar terms as it captures a *lacing together*, not entanglement or weaving that keeps threads ontologically intact, but an enmeshment and subsumption one an entity with a course-changing process that draws in the entities.
2 Metaphysical bodies include any entity that is capable of affection, transformation, and/or production.

3 Here, speaking of a table would not even be appropriate, as it invoke thinking the object as performing a functional unity, whilst the object is not compulsively engaged with on this function, but on its sharp tip. This critique is made possible through utilising Bergsonian images. Therefore, from this point onwards, objects as enlaced in compulsive configurations are either invoked as objects or extracorporeal materiality.

4 This builds on the experience of 'getting stuck' in a situation or activity, which is explored further in Chapter 3.

5 In the Netherlands, coffeeshops are places where citizens over age 18 can legally buy cannabis for private consumption. Sales points originated in regular cafes where cannabis sales were legally tolerated, hence the name.

6 Chapter 8 puts forward how this is not without controversy in light of the growing neurodiversity movements and autism advocates challenging of behavioural therapy as applied to 'treat' autism.

References

Andrews, G.J. et al. 2012. Moving beyond walkability: On the potential of health geography *Social Science & Medicine* 70(11), pp. 1925–1932.

Ash, J. and Simpson, P. 2016. Geography and post-phenomenology. *Progress in Human Geography* 40(1), pp. 48–66.

Bissell, D. 2009. Obdurate pains, transient intensities: Affect and the chronically pained body. *Environment and Planning A* 41(4), pp. 911–928.

Bliss, J. 1980. Sensory experiences of Gilles de la Tourette syndrome. *Archives of General Psychiatry* 37, pp. 1343–1347.

Braidotti, R. 2011. *Nomadic Subjects: Embodiment and Sexual Difference in Contemporary Feminist Theory*. New York, NY: Columbia University Press.

Chouinard, V., Hall, E. and Wilton, R. eds. 2010. *Towards Enabling Geographies: Disabled Bodies and Minds in Society and Space*. Farnham: Ashgate.

Cowhey, S.P. 2005. *Going Through the Motions: Coping With Autism*. Baltimore, MD: Publish America.

Davidson, J., 2010. 'It cuts both ways': A relational approach to access and accommodation for autism. *Social Science and Medicine* 70, pp. 305–312.

Deleuze, G. 1991 [1988]. *Bergsonism*. New York, NY: Zone Books.

Deleuze, G. and Guattari, F. 2004 [1980]. *A Thousand Plateaus. Capitalism and Schizophrenia*. London: Continuum.

Dewsbury, J.D. 2003. Witnessing space: 'Knowledge without contemplation'. *Environment and Planning A* 35, pp. 1907–1932.

Duff, C. 2011. Networks, resources and agencies: On the character and production of enabling places. *Health and Place* 17(1), pp. 149–156.

Gleeson, B. 1999. *Geographies of Disability*. London and New York, NY: Routledge.

Hall, E. and Wilton, R. 2017. Towards a relational geography of disability. *Progress in Human Geography* 41, pp. 727–744.

Hansen, N. and Philo, C. 2007. The normalcy of doing things differently: Bodies, spaces and disability geography. *Tijdschrift voor economische en sociale geografie* 98(4), pp. 493–506.

Harrison, P. 2009. In the absence of practice. *Environment and Planning D: Society and Space* 27, pp. 987–1009

Lawson, W. 2005. *Life Behind Glass: A Personal Account of Autism Spectrum Disorder*. London and Philadelphia, PA: Jessica Kingsley Publishers.

Lea, J. 2009. Post-phenomenological geographies. In: Kitchen, R. and Thrift, N. eds. *International Encyclopaedia of Human Geography*. London: Elsevier, pp. 373–378.

Leder, D. 1990. *The Absent Body*. Chicago, IL: University of Chicago Press.

Levinas, E. 1998. *Discovering Existence With Husserl*. Evanston, IL: Northwestern University Press.

Lingis, A. 1998. *The Imperative*. Bloomington, IN: Indiana University Press.

Manning, E. and Massumi, B. 2014. *Thought in the Act: Passages in the Ecology of Experience*. Minneapolis, MN: University of Minnesota Press.

McCormack, D.P. 2017. The circumstances of post-phenomenological life worlds. *Transactions of the Institute of British Geographers* 42(1), pp. 2–13.

Murphy, R. 1995. Sociology as if nature did not matter: an ecological critique. *The British Journal of Sociology* 46(4), pp. 688–707.

Philo, C. 2007. A vitally human medical geography? Introducing Georges Canguilhem to geographers. *New Zealand Geographer* 63, pp. 82–96.

Schillmeier, M.W.J. 2010. *Rethinking Disability. Bodies, Senses and Things*. New York, NY: Routledge.

Wylie, J. 2006. Depths and folds: On landscape and the gazing subject. *Environment and Planning D: Society and Space* 24, pp. 519–535.

Wylie, J. 2009. Landscape, absence and the geographies of love. *Transactions of the Institute of British Geographers* 34, pp. 275–289.

Wylie, J. 2010. Non-representational subjects? In: Anderson, B. and Harrison, P. eds. 2010. *Taking-Place: Non-Representational Theories and Geography*. Farnham: Ashgate. pp. 99–114.

6   Object excess

Movement in concert

The previous chapters demonstrate that compulsions do not start with sensations and/or with not-just-rightness, rather, they start with bodily configurations towards the extracorporeal that are often accompanied by sensations and not-just-right feelings. Following Ash and Simpson (2016: 55), attuning to the body-world relation that underpins compulsivity, allows placing:

> ... an emphasis on the vitality of embodied experience, on the dynamism of felt intensities that find corporeal expression in the feeling body, and also an emphasis on the ways in which the body-subject undergoes constant processes of 'affectual composition' in and through its relations with a material-agential world.
>
> (Ash & Simpson, Geography and post-phenomenology, 2016, p. 55)

Hence, by looking at the vitality of the extracorporeal we can better pinpoint how certain movements come about in response to the non-human extracorporeal. Therefore, the body-world relationship needs to be repositioned in accordance with Lingis' environmental imperative. This chapter explores how the body becomes configured by focussing on its socio-material situation, in particular, on how objects and spaces come to appear in the compulsive interaction. The term 'objects' denotes a metaphysical object or body; the term therefore includes not only 'inanimate' 'things' but also non-human animals and plants.

The broader extracorporeal world plays a curious role in compulsivity. In current medical and clinical understandings, objects have been dismissed as randomly involved or as merely available when biological impulses urged people to perform a compulsion. Similarly, in psychoanalytical analyses of compulsivity, objects emerge as representations of desirous relations of the self or others, but they are not considered on the basis of their materiality and form. As I argue in Chapter 2, aligned with the analytical focus on the body and the human, objects have not only been pacified in analyses of compulsions, but they have been rendered forgettable. For good seasons perhaps, as objects are not part of the biological body, nor do they suffer. In addition, they exist in unlimited forms, sizes,

DOI: 10.4324/9781003109921-7

textures, temperatures, colours, and weight, which makes their inclusion in positivist analyses of compulsivity equally limitless. That would make it seem as if any effort to account for this analytical category is futile. This chapter disputes this.

In the same ways as compulsive subjectification, processes mark the shift in embodiment of wanted, purposive life and tasks, what becomes of objects, spaces, worlds can be traced in similar ways. What follows is the construction of the extracorporeal participation in the configurational theory offered in the previous chapter, and in turn, trace how compulsions re-create objects, materiality, and re-assemble life worlds. The analysis first explores what objects become involved in compulsive interactions, deconstructs objects to work through how their involvement features in the compulsive configuration, and reconstructs the extracorporeal world in accordance with processes of perception and embodiment. The chapter ends with denoting how configurations and affirmative compulsive interactions accumulate spatially.

Objects in compulsions

Compulsive interactions reveal a lively, active, and sometimes demanding and violent world that remains largely unthought, as their emergence in compulsive interactions differs strongly from other-than-compulsive interactions. The anticipative landscape of tension from which compulsive interactions emerge is in part shaped and regulated by material objects that announce themselves in the compulsive configuration. For instance, Dylan reflects how he experiences object presence during compulsive moments:

> Dylan: I often have this thing that during intense tic moments I want to grab hold of something; a chair or a bottle or a ehm ... pillow or whatever.
>
> DB: Grabbing hold to avoid toppling over?
>
> Dylan: No. No, often just because it exerts power over me or something ... it feels like ... that I just really have to grab hold of something.

It is a striking account of the vividness of the materiality around him and the dominating force certain objects can acquire. In these moments, his body violently configures towards that chair, bottle, or pillow, and he feels almost attacked by an object and must attend to it through interacting with it, whilst enduring other tics. Certain objects exerting power over a body are also familiar to Nora. Iron pillars tend to acquire a ferocious vitality when she gets near a particular set:

> Nora: Along the iron pillars, that's a must, I really have to! It [touching one with her chin] is – within quotation marks – a habit, but at choir practice I really must do it.

DB: And what do you have to do with the pillars?

Nora: Sliding my chin along it

DB: Like, vertically?

Nora: Yeah, like feeling with my chin, like so *demonstrates on her coffee cup* I know what material it is, I know it's cold, and still my chin has to slide over it. These pillars have such an attractive forcefulness over me. I just must do it.

DB: Okay, and do you remember that thing before you see it?

Nora: Oh yes, and when I'm stood near that pillar, I'm like 'come to me, come to me' *smiles* and then I think 'no, don't, because you'll look like a fool', and seconds later I find myself sliding it along the thing! *laughs*

Every time she is in the vicinity of such a pillar, its overwhelming evocativeness can hardly be negated. What she experiences is distinct from other kinds of pillars, the forcefulness differing compared to other objects she compulsively interacts with. Whilst iron pillars were not remarked by other participants, other objects did seem to be more prone to emerge within compulsive interactions, for instance common household items like glasses, mugs, and socks. Not one object took part in compulsions by all participants, and, in turn, most objects being interacted with compulsively at one occasion did not necessarily evoke compulsions at another occasion. When asked if his compulsive interactions include objects that he feels particularly drawn to, Joe responds:

No, no, no, it's about what I think I need to do, but not anything specifically, no. (...) Also in shapes, I'm not picky. That's with everything, seeing something in proportions, I don't care, I'm happy with everything; it's not conscious like 'I have to feel that shape or that shape'.

Indeed, compulsive interactions are performed with a wide variety of object shapes; generalisations of the shape and texture of objects cannot readily be made. For instance, the study on which the book is based registered mugs having been rubbed and pinched in different places, placed on a shelf, and (re)ordered in relation to shelves and other mugs. Objects such as soup bowls, glasses, and stone statues reappear through their flat undersides when meeting cupboard and showcase shelves. In analytical terms, whilst object and body assemblages, such as long drink glasses on a shelf in the cupboard at eye-level, precondition compulsions, material assemblages *as such* cannot explain how objects feature in compulsive interactions, nor why at certain moments worlds turn compulsive. Considering what parts, aspects, and dimensions of objects compulsive configurations revolve around provides insights that are more helpful.

Similar to how bodies shift into compulsive enlacements, objects and materials shift into another appearance when they are compulsively inter-acted with, before they configure back to a other-than-compulsive appear-ance. Properties of objects that articulate with functionality (as table, as jeans, as pencil), aesthetics (as red, as porcelain, as having neat edges), and meaning (as gifted by a departed friend, as childhood memorabilia, as marking a personal achievement) play a very small role in compulsive engagement, and cannot determine whether an object will become enlaced in a compulsive act. During the interview, Bill occasionally picks up his water bottle in front of him to take a sip, and every time before putting it down, the round ball-like bulges on the top and bottom of the bottle need to be pressed in hard with his thumbs. In these moments, the bottle acquires an appearance of press-able, smooth plastic 'ballness'. Later on, this appearance configures similarly, but also through the bulging lower part of the bottle. Whilst plastic often returned in the registered compul-sions in the current study, other materiality that featured in compulsive interactions included varieties of glass, metal, paper and cardboard, fab-ric, wood, soil, plants, and animal fur. Compulsively interacting with cer-tain materials can be less annoying than with others, and sometimes this is even mildly enjoyable: material such as cold marble has an 'attractive forcefulness' for Sage, for no other reason that it feels like a cold sleek, smooth hard surface.

Craft hobbies, like woodwork and leatherwork that involve ongoing interactions with special object materials, can create situations that blur the distinctions between compulsive and other-than-compulsive action. The 10 × 15 cm thumb-thick square made of soap-stiffened raw wool Mina worked on, had been intended to become an artwork for a friend to hang on the wall. Whilst pressing the special needle in the wool, putting new wool tufts on or tearing tufts off, she narrates how the piece and its details generally – to use her words – "improve, but more often worsen", as com-pulsive sensibilities strongly guide her artistic standards. In the process, the piece invariably reappears as stacked layers of wool, a composition of three figures, and a smooth then bumpy then bulging rough surface[1]. As such, it reconfigures between an emerging artwork and a lively form bursting with just-rightness and not-just-rightness. With compulsive and other-than-compulsive interactions, the dimensions of objects acquire a dynamism, which leads to a succession of rapid creations and collapses of compulsive situations. Compulsivity then exposes a certain instability of the material environment; as if compulsions constitute a glitch in an otherwise self-affirmative and more or less coherent world.

Non-human objects thus do not enter compulsive interactions as whole and complete things. Rather, the becoming-compulsive of a situation striates them according to their capacities to take part in that precise interaction that the configuration works towards. These capacities are, however, not only governed by the object itself, but by a combination of

certain forms of presence that create an anticipative tension. These capacities also emerge in accordance with the physicality and sensory registers
of the human body, which is determined in processes of perception. For
instance, Sara always carries with her two small stuffed animals when she
leaves the house. She takes them out of her bag on a train journey back
home after a therapy session and rolls between her thumb and index finger whilst staring out of the window. For a moment, it stops being the
small stuffed animal and reappears as fabric of the right texture with a
movable lumpen thread. Another compulsive transition occurs with Elisa.
Immediately after she uses the steering wheel in her car to drive her and
myself to her house, she leans forward and twice presses her chin onto it.
From its 'objectness'[2] as a steering wheel and its functionality to steer the
car during the drive, the steering wheel 'falls apart' and is addressed on
its hard, leather, patterned, curved closeness to Elisa's body. Elisa's hands
had been exposed to the steering wheel during the trip, but when she no
longer needed it for driving, its sudden demand to locate against the skin
of her chin then revealed a different appearance. In ontological terms,
during the configurations towards compulsive and other-than-compulsive
situations, objects thus retain an autonomous existence outside their
appearance in perception. However, it remains unclear if particular compulsions can be produced by virtue of introducing certain objects to a
certain situation.

Aside from texture and form, varieties in wetness or dampness also
emerge as evocative, as Sage presses the damp fabric of a white sock into
her left index finger before putting it on the line. On another occasion, a
washcloth is brought to her upper lip and nose when she picks it up from
the laundry basket. Following temperature difference, saucepans reappear
as scorching hot round metal formations, and in terms of colour deviations,
tomato stems in the sink shift into "black bits" in a silver-grey sink that
therefore had to be removed. Every object that a person encounters thus
introduces an intricate collection of materials, temperatures, textures, colours, and shapes into life worlds that have a distinct set of many possibilities
to emerge in compulsive interactions. Similar to David Hockney's painting
'Chair', which is a constellation of multiple perspectives on a single object,
visualises how objects appear differently in terms of their colour, texture,
materiality, and temperature from different angles. They remain recognisable as whole, unified objects that also retain a sense of functionality, but
are simultaneously recreated in their reconfigurations towards multiple
compulsions.

Objects tend to be more prone to become involved in compulsive interactions because of their incapacities to endure certain violence through tearing, rupturing, or breaking. For instance, paper was pierced by charcoal,
dandelions were kicked, fineliners had their tip split, and tips of fountain
pen nibs were split. In turn, compulsions also seem to follow their capacities to bestow violence on the human body through their materiality, form,

and temperature. These actions involved metal pins pressed hard into index fingers, closing doors that were closing trapped and pressed thumbs, metal pans lifted up from the stove burning chins, and many more. Indeed, often objects do not just need to be 'touched' but "pressed hard". Accordingly, for Sage this makes knives "vicious". In fact, objects' capacities for violence were a crucial in compulsive interactions Rhys performs:

> It's more that the glass can shatter, with the hot thing [oven dish] it's that it's hot, so it doesn't matter that it's made of iron, or is an oven dish or whatever. It's just the fact that it's hot that makes it interesting, that the glass can break, the fact that the door can close is what makes it interesting.

Mina agrees with Rhys that object "fragility" and "brittleness" then becomes "a provocation", which must be pushed "until it's completely broken, otherwise it's feeble". Interactions with such objects even go on to mutilate her body as "it really can't be stopped". For instance, biting down on toothpicks stuck vertically between her jaws softly "doesn't count, as you really have to reach that maximum point of forcefulness", because "if it just cracks lamely, then you're not satisfied (...) if it doesn't crack painfully". Some of Mina's experiences closely resemble those of a person treated by Meige and Feindel (1907 [1902], np):

> The patient could not keep a pencil or a wooden penholder longer than twenty-four hours without gnawing it from one end to the other. The same thing happened with the handles of sticks and umbrellas; he destroyed an extraordinary amount of; them. To help him out of this predicament he was seized with the idea of having metal penholders and sticks with silver knobs. The result was most disastrous; he bit at them all the more and as he could not destroy the iron and silver he very soon broke all his teeth.
>
> (Meige & Feindel, *Tics and their Treatment*, 1907 [1902], np)

What stands out is how in such compulsive entanglements with dangerous objects, the compulsive experience is incredibly intense. Objects become intimately known through the testing of their capacities to damage the body or be destroyed by the body. In such compulsions, one is not only at once a body (after Leder 1990), but becomes the skin, its bouncy fleshiness, as well as the metal sharpness of pins, and the weight and pressure of their conjoining. In such a compulsive configuration the pin's metal sharpness reverberates in muscles, and an upper lip becomes the soft roundness of a statue's head when it becomes pressed into it. In the moment the body and object meet; when they express a 'punctuation' as Bliss (1980) puts it, there is a concentrated movement towards each other. When the configuration completes in the moment of a climactic just-rightness, the body and object re-form in each other's image. Hence, certain knives cannot be handled without causing damage, because the punctuation expressed by the form of the knife punctures the skin. Both body and object undergo a reformation that is ordered by that particular just-rightness until the intensity wanes and other-than-compulsive life resumes.

One way of thinking about destructive compulsions that mutilate parts of the body and objects is that it refrains from invoking future interactions. When all toothpicks have been cracked, no more biting compulsions are needed, adding a certain future in which these interactions are no longer possible. Destructive compulsions can also turn on the body, as Mina painfully broke four of her molars on the left side of her jaw as result of compulsive biting interactions. Years of having a 'chew compulsion', as she terms it, had strengthened her left-side jaw muscles. This, as well as the effects of Selective Serotonin Reuptake Inhibitors (SSRIs)[3] she had been taking, satisfying the urge culminated in the destruction of her body. She had not had her molars replaced at the time of the research and in the past she had broken fillings in her molars: with her jaw muscle strength, she simply breaks any dental implant, which refrains her from compulsive chewing but presents problems in other-than-compulsive life. This compulsion reflects many other compulsions that are painful and destructive for pushing bodies to and past their limits. Gum was chewed to offset strong urges to bite down hard and a mouthpiece was worn to cushion Dylan's teeth. Indeed, not all violent compulsive situations come to fruition. The research participants argue that otherwise tomatoes and other soft fruit and vegetables would have holes pressed in them, thin glass would be crushed between molars, paintings would be felt, scratched and poked holes in, and glass tables would be smashed with hammers. In addition to alterations to the processes of perception and desensitisation of the person involved, objects themselves can be altered or re-placed; tomatoes placed in a bowl with other soft fruit and vegetables, thin glass cups were replaced by thick glass cups, and paintings were guarded and made unreachable, glass tables were replaced with wooden ones.

Excess collections

The involvement of objects in compulsive interactions cannot be discerned from looking at their properties, and they consist of collections of materiality, texture, temperature, etc., that shape their possibilities for partaking in compulsive situations in the image of the body. This section examines what makes them 'stand out' amongst others that 'earns' them a place in the compulsion. In other words, of all the objects available to people at all times; why does *this* compulsion take place with *this* object? That does not mean that there is a sense of competition between objects that would underpin this, however, there is a visual liveliness that emerges from object collections. Such liveliness emerges when objects do not appear distinctly different from each other, losing their boundedness and 'individuality'[4], and are then perceived to 'overflow into each other' in several ways. Ginny, for instance, recalls such experiences of distortions in a large furniture shop and supermarket:

> Look, like in this space all is standing still, but when I go to such a supermarket, and I've just been to [large furniture shop brand]. Then I see so many new things, but it seems like everything is moving, despite it not actually moving. But it demands all attention – or something – from me. Also, when I don't give all of it attention and I'm trying to look at one thing, then I see vermicelli, and then it acquires wriggling, glowing sides, to put it like that. I see all kinds of thingies *makes crawling motion with her hands in front of her face*. That's how it is, I can't explain it in other ways. (...) When I'm in a supermarket – imagine you're in front of a new stand of shelves – and at that moment I see everything that's on display, but I also don't know what it is actually, but I do see all of it. And that kind of hurts my eyes. Then it's as if all those colours, all those thingies, all those little patches just enter me, I don't know what it is, it just doesn't happen with me. Someone else gets the information immediately and sees a bag of liquorice, but I just get stimuli[5], and nothing happens. (...) It's a kind of work of art, blocks with all kinds of colours which also move the whole time. Then I'm just standing there like *stops mid-movement and swipes an imaginary stand of shelves with her eyes wide open* (...) Six to seven boxes is then one block of colour. Then somewhere I see a box, and thereafter those colour blocks again. (...) Another person sees a box of vermicelli, to put it like that, and it's always the same box of vermicelli. For me it's an entirely new box of vermicelli every time I encounter one.

Individual objects thus do not seem to 'stand out' from other objects. In fact, in the experiences Ginny describes, it is the other way around; it is in their collective appearance that objects' vitality seemingly intensifies. In their unstable presence, they take up more space with their vibrating movement and their colours multiply beyond their material boundaries. It enlarges them, not just in terms of taking up more of people's field of vision, but also in the sense that they demand more attention, as Ginny argues.

Furniture shops and supermarkets both consist of many finely detailed objects, whilst also tending to be noisy. The objects vibrating in Ginny's perception makes them difficult to distinguish from each other, which presents the surroundings as a dazzling world in which it is challenging to orientate oneself. Landscapes of tension from which compulsions erupt and that largely remain felt and intuited can thus also become perceptible visually. While such experiences are not necessarily immediately followed up by compulsive interactions with the moving objects, they can be regarded as configurations that capture the human, as these situations stop Ginny in her track leaving her confused and unable to continue other-than-compulsive life. Negotiating 'becoming stuck' in busy environments by moving through them quickly, can only work if only glancing at one's surroundings suffices in continuing other-than-compulsive life.

The shifting appearances of unstable objects may indicate fast-changing configurations, that call attention to different objects and require different engagement. This expresses as a rapid attunement to textures, forms, and colours that quickly alter formation that produces a rhythm so intense that it suspends assimilation of meaning from a selection of this. Consequently, bodies constantly require to (re)configure to the next, and the next, and the next non-human and cannot 'keep up'. This disorientation leads Ginny unable to "feel the distance from [her] body to a thing". Undergoing this experience standing still in one place is already difficult and demanding, but it is intensified by moving around. It is an inexhaustible situation of reconfigurations, that in the less busy context of her home, walking from the kitchen to the sofa in the living room with a mug of tea invites a series of configurations. Her body configures towards, and quickly needs to be reconfigured from, a warm roundness in her hands, to trailing the path from the hip-high cold metal flatness towards the knee-high softness, towards the sharp hard pointed edge, to the barking low behind her, to the small colourful patches and curves in wooden frames at eye-height on the right, to the rectangle of colourful round shapes at hip-height, with the horizontal and vertical lines overlapping with every step, ever intertwined with fabric clenching skin that pulls and pushes and rearranges the little hairs of her arms with every next movement. Feeling less intense in the less busy context of her home, she explains that "if I don't see that the table is close-by I can't feel it" and she "walks into it", "makes a lot of noise when doing the dishes", and "plumps down on the sofa", because she "cannot make that judgement".

The shifting appearances of objects as collective blocks of colour, presenting as fuzzy forms with unstable boundaries is a familiar phenomenon in autistic experiences. Following theorisations of autistic perceptions helps to clarify how the extracorporeal world changes appearance. Indeed, autistic experiences of busy environments, such as supermarkets, can equally overwhelm people with autism, which feels as 'senses being shut down' and being overcome with a 'meaningdeafness' (Williams 2005, Davidson and Henderson 2010). Based on the autobiographical work of autistic authors to better understand autistic perception vis-à-vis the other-than-autistic perception, Erin Manning and Brian Massumi (2014) argue that objects differ in their manner of presence. Where non-autistic perceptions of a certain formation of colours, textures, volume, and weight immediately become a flower that can be smelled, jumping to seeing objects almost exclusively in terms of usage, autistic perceptions are characterised by the introduction of infinite possibilities of this formation that do not prioritise use value. In such perceptions, autistic people first and foremost see the 'fielding' of a flower in its own right; the subtle movements of its leaves in the wind, the brightness of its yellow petals, and the light bouncing off the little soft hairs of its skin. The flower *also* has a scent, and one of the infinite options of this affective fielding is for a person to smell it. The flower exists in its own right and has many affective qualities, but it does not exist only *for* the human. The flower always already exists for all its affective surroundings in the autistic perception, which is thus a much more complex picture that is much less straightforward to comprehend.

These understandings chime with Ginny's experience of perceiving a formation of colours, textures, and lines, but not seeing this formation immediately as a box of vermicelli that she can pick off the shelf and put in her shopping basket. Manning and Massumi citing Etienne Souriau (2009) and Gilbert Simondon (2005) then explain that perceptions of objects' fielding render sensible how objects exist in resonance with their surroundings, and, in effect, how their appearance is formed in accordance with others. Indeed, Ginny's experience of the colour-texture-line formations not always shifting and often not staying in vermicelli boxes on a shelf is testament to that. Another compulsive interaction that suggests relations of appearance between objects is the compulsive placement of two plant pots in accordance with the tiles of the windowsill, a small Buddhist figurine, and a round bush outside in the garden. The interaction works towards a just-right alignment of the side of the plant pots, in accordance with the horizontal centre of the figurine, of which its underside needs to cross the line between the tiles and in the middle between the edge of the windowsill and the frame of the glass, and of which its head needs to be in the centre of the round outline of the bush, as seen from straight ahead. Getting this just-right ensued shuffling of the figurine and plant pots. However, what this compulsive interaction works towards differs:

> Ginny: It just depends on what kind of itch I get *laughs*, because when I get the 'tile-itch' then the figurine needs to be in the centre of the tile, and if I notice the space between these pots, those pots need to be equally distanced on the windowsill.
>
> DB: So it's kind of ...
>
> Ginny: ... the first thing you see, yeah, and what kind of itch you get from it.
>
> DB: So it's related to what you see?
>
> Ginny: Yeah. Yeah, yeah, it's very much what I'm looking at.

In this instance and many others that entail ordering and alignment, visual perception of objects in formation determines what elements and dimensions are the organising principles that align other object by virtue of their dimensions that are produced in and with the just-rightness, or 'tile-itch', as Ginny puts it. In compulsivity, objects are thus not perceived *qua objects*, rather, they exist *only* in excess of themselves, and towards and amongst other objects. Indeed, organised by their inscription of movement towards both human and non-human bodies, singular objects only exist in their self-referential excessiveness.

Compulsive interactions are underpinned by several principles of object excess, one of which is size. Objects of similar size induce easier pattern recognition and demand similar bodily engagement, for instance, books stacked in piles as high as they are wide need to be placed equally distanced in a standing, laying, standing, laying, and laying formation on Elisa's windowsill. Another principle of excess that underpins compulsive interactions is objects sharing a dimension. For instance, Alan aligns two long and narrow wooden slats on and with the edge of a long rough wooden desk in his garage. The length of the slats become affective and configured in accordance with the desk edge, as the length of the edge appears similar in length because of surrounding objects. The dimensions that become apparent in compulsive interactions can change with the new post-interaction situation, which can incite a new compulsive configuration. Ginny closing a cabinet drawer with two hands on either side of the drawer aligns the cabinet with the line on the floor through which it moves away from the table it rested against. After having circled her index finger around the handles of the drawer, she needs to move the cabinet back against the table. In this series of interactions, the cabinet drawer configures with the cabinet, which configures with the floor and becomes reconfigured with the table.

As we had established that compulsive interactions reveal objects as collections of lines, textures, temperatures, colours, etc, fielding in compulsivity then takes place with these collections themselves. Configurations towards compulsive situations see objects cease existence for the human and re-emerge as a collection of self-luminous surfaces. This clearly presents problems for Ginny, as every time she encounters a box of vermicelli it takes effort to retrieve it from its *collection in excess* and make sense of it in human other-than-compulsive terms. However, in compulsive situations, object formations can regain individuality and recognisability by them emerging as specific rhythms of self-luminous surfaces. In turn, compulsive interactions become affirmative performances of perfecting the rhythm. Compulsive situations thus manifest a reformation of the extracorporeal into new material communities that cultivate more-than-human cultures.

Movement with

As the previous section and chapters have stated, objects can "pull out" a certain, possibly anticipated – but unthought – engagement from the human body, which takes place in conjunction with objects' collective appearance. Whilst in this ontology so far, the human body is rendered submissive in and to compulsive situations, it retains an organisational capacity in how compulsions are performed that does not hinge on notions of reason, free will, or social contexts.

Compulsive interactions involve other people, or, more accurately put, other human bodies. It is not the 'humanity' of other people that is affirmed in compulsions but particularly compelling fabrics of clothes, such as fleece for Elisa, or particularly compelling body forms, textures, temperatures, and colour. Indeed, other human bodies seem to become purely involved on their materiality and form, not on their social dimension, even when the compulsion resembles meaningful social interactions. For instance, after a co-worker who had had reconstructive surgery after a mastectomy, Elisa "just wants to touch her breasts" because she "just want to know how it feels"; not because of the sexualised connotations of breasts: "No, they could as well be her knees (...) it's not symmetrical, the one is bigger than the other, and that has to be compared". The difference between the shapes of these 'objects' articulate a not-just-rightness that captures Elisa, which she can only negotiate by feeling these breasts.

In a precise mirroring of non-human objects, the body itself thus becomes perceived as self-referential and excessive form in compulsive moments. That does not mean that the body is de-animated in compulsivity but that it reflects the animation of the extracorporeal in compulsive situations. Whereas compulsive ordering and/or aligning of objects and touching objects seem like categorically different kinds of compulsions, they happen according to the same principle. It is more like an alignment of the body form with objects; the tip of the finger is soft and cushiony in the excesses of a sharp point. Other, softer material is not interesting enough because it cannot be determined exactly how the interaction takes place, and the body is very often in sight. There is a visual connection between a texture and the hand, in addition to the hand's capacity for sensation and determining in detail how the interaction takes place, which is also true for the lips and underside of the feet. When Joe compulsively touches animal fur, bark, or a painting, he argues that the act ensures this is the material form that it purports to be. Joe contends having to stroke taxidermied animals in museums: "then I just know that such a beast has super soft fur, and then I just have to touch it, that's just how it is". During a compulsive moment, the interaction is a bringing together of materials; sensation performs a guidance to the intensity of the meeting of the materials; the meeting and alteration of form. In the compulsive stroking interaction, the soft and sensitive cushions on Joe's hand meet the soft and bouncy fluffiness of the animal fur with the light gliding pressure that emulates these shared qualities.

Echoing Hoel and Carusi (2015: 78) in their analysis of the body in Maurice Merleau-Ponty's phenomenology, the bodily flesh articulates as "operative, organizing force", which in compulsive interactions seems to be amplified. In turn, compulsivity demonstrates how the flesh produces organising principles of the corporeal with the extracorporeal. The flesh as organising force for the body and its constituencies also takes place outside the human body and follows complex, pre-personal, and anonymous sensibilities. When Ginny walks past a banana, it suddenly emerges as necessitating compulsive repositioning among other fruit in the fruit bowl. She picks it up and places it back at an angle because:

> It needed more space, kinda, otherwise I'd be sad for it. He'd be more comfortable, yes! *we both laugh* Yeah, I'm just happy for that banana! Yeah, I really have that with these things, and the reward is then a reason for, I think, that they would like it better if they, yeah, don't lie against the other one *laughs*

Extended from the body, such fleshy sensibilities that conjure up the particular collections of object-excess put forward an intricate understanding of the sensorium of the banana amidst the other fruit in the bowl. It purports a detailed banana-shaped constellation of the sensations of textures rubbing off on each other and the pressures of weight on its peel all relative to the positions of the banana in the bowl. The position that makes the situation just-right again then hinges on the most 'comfortable' position of the banana in the bowl. Similar to Ginny's experiences of compulsive tendencies in the supermarket, these extracorporeal sensations emplaced in the banana have a pertinent immediacy. She can look at something for a brief moment and then has to look away to avoid the urge from coming up. When she looks at that fruit bowl on a different occasion, she narrates:

> If I keep looking at it, I have to look away immediately, because if I don't ... *she gets up and walks to the fruit bowl* Then I have to position it like this ... *moves fruit in bowl* because this has a dent in this, which is noticeably different, because if it's positioned like this ... *moves fruit in bowl*. The thing is, when I keep looking at it, it's not right, so then I have to put the dent and the ridge here *points towards other fruit* of out of sight, because otherwise I find that very annoying.

With the speed at which the banana in the fruit bowl configures towards a situation of not-just-right textures, touching with too much pressure behind it is breathtaking. Arguably, in particular the banana has such a strong appeal to interact with because it has a velvety soft skin, not too dissimilar to human skin. However, rather than to make a turn to anthropomorphising objects, we are reminded of Merleau-Ponty's (2013 [1945]) assertion that worlds are created in the interrogative mode. For Ginny, the banana could require immediate compulsive engagement because its synaesthetic textures are simply more easily recognisable.

Alphonso Lingis' theories of extracorporeal, distributed imperatives (1998) takes this further by arguing that objects, spaces, humans, and non-human animals that 'make' worlds evoke human movement and engagement in accordance with them (2000: 29):

> Our movements are not spontaneous initiatives launched against masses of inertia; we move in an environment of air currents, rustling trees, and animate bodies. Our movements are stirred by the coursing of our blood, the pulse of the wind, the reedy rhythms of the cicadas in the autumn trees, the whir of passing cars, the bounding of squirrels, and the tense, poised pause of deer. The speeds, slowness, and turns of our movements come from movements we meet about us.
>
> (Lingis, *Dangerous Emotions*, 2000, p. 29)

In this sense, the corporeal moves with and according to the extracorporeal (see Wylie 2006, Anderson and Wylie 2009). As worlds 'conduct' bodily movement, the excessiveness of objects continually reorientates bodies, with compulsions articulating such orientations through configurations of the body outside movements that induce formations of representation from which meaning and reason are derived. Indeed, if compulsivity can be thought to emerge from the body's movement with objects, compulsive interactions suddenly become more understandable. Ginny's compulsive ordering of dispersed cans closer to each other highlights their distinctive shape, shine, temperature, and texture amongst other objects:

> I don't like putting something in just some spot, because I'm like 'this is what they like' *smiles* 'this is comfortable for them. I was also delighted that I had them [cans] all there (...) Yeah, and I'm quite happy that I have that whole family together, those cans.

Compulsive interactions fold collections of object-excess onto and into each other, recalling the permanence, or reifying the temporary forms and textures as felt *through* each other. Placing a glass on the shelf in a kitchen cupboard moves the flatness of the upside of the shelf in concerted flatness to the underside of a glass, as well as the rigidity and musicality of the glass when it bounces off another glass that had been placed there already by Bill. The concerted movement of the same size and roundness of the coins is met and amplified in the figure recovered through their place in the brown sand outside the back door to the garden when Ginny steps over them. The texture of the dry, stiff fabric of the two socks moves with the movement of the two other socks still on the line of a laundry rack when they are made to touch by Cai. The increasingly grainy texture of the paper moves with the dry, velvety charcoal stick when it glides over it again and again and again by Mina.

Similar to object movement captured in their excess, human bodies change appearance towards and in accordance with the extracorporeal world. Differences in interacting with corners and pointed edges as well as surfaces, temperatures, and textures produce specific sensations that affirm the body's movement; to be bouncy here, soft there, and rigid there, to arch here, have an edge there, and a bump there. Compulsive configurations then fold the human back onto and into itself during the interaction, as warm when touching something cold, as soft when stroking the grain of rough wood, the rigidness of a finger's bone when pressing the tip straight into a mirror, the permeability of one's skin when pressing into a knife's tip.

With situations changing because people walk around, lights are turned off, the sun comes out, other people join or put something on the table, houseplants shed leaves, another other-than-compulsive activity is embarked on, new compulsive configurations can take place, as human bodies and objects acquire new dimensions, and as such, new appearances. For instance, when Bill sits in his chair he compulsively interacts with a water bottle in front of him, and standing up from his chair, he suddenly needs to perform multiple other compulsions with other elements in his surroundings. What did not stand out as a distinctive corner or had a certain evocative texture from the position of the chair, may do so from the angle of standing up, with the corner suddenly visible and the light bouncing off a pillow differently. Equally so, what may have configured Bill's body towards a compulsion can have no longer demanded compulsive interaction.

Whether or not object dimensions evoke a compulsive interaction and if so in what way that would happen seems to follow certain broad principles. These principles follow the fielding of the body itself as directed by its infinite capacities to move and ways to interact, and seem to form a range of possibilities for compulsive configurations to happen. One such principle relates to bodily reach, as Sage ponders:

> I think if it [interactions] can't be done, then it doesn't happen, maybe I register it, but that it immediately vanishes indeed (...) It's more the things that I see around me, and sometimes, when I'm in the classroom I look at the wall and then I think 'ooh, I must touch that for a second' but it's always one that's close to me, so within range, even though I'd have to slide my chair backward, but I can reach it. It doesn't happen [like that] at great distance I think.

This is also the case for Lowri, as she explains that when things are out of reach she does not feel compelled to interact with them. Moreover, remaining configured towards certain objects offsets compulsive configurations to take place. As Dylan argues, "it's often also that having something in your hands often ehm ... also helps with walking". Potentially, the ongoing configuration of his hands prevents other compulsive configurations to take place, and indeed, rarely are hands freed to compulsively interact with objects within reach in passing.

Certain elements at greater distance can keep bodies compulsively configured when they involve being looked at. For instance, a small red light on the DVD player requires Mina to 'reciprocate' its piercing articulation by staring at it, which prevents her from reading. To configure to reading, she needs to get up from the sofa, turn the player off entirely, and find a new, uncomfortable position on the sofa to also face away from it. Another shaping of the compulsive situation through the fielding of the body entails the extension of the biological body into objects yielded for particular other-than-compulsive tasks. The front wheel of Sara's scooter needs to align with the long stripes painted on the tarmac when she drives on and off elevated junctions. Similar to glasses needing to touch each other when being put in the cupboard, this compulsive interaction entails a movement of Sara's embodied scooter with the 'vertical' line the stripe induces. As the scooter reverberates Sara's sensibilities through its materiality in constitution of the compulsive situation, the non-human resonates in and through the human. Thus, in the compulsive moment the ontological difference between Sara and the scooter dissolves.

Compulsive situations that are similar in the ways they are instigated and in the interaction it takes to reconfigure to other-than-compulsive life can differ in their volatility. Alan, for instance, walks with a crutch that becomes included in a type of compulsion that required him to place its end in the centre of a bounded surface area, such as a white piece of plastic on a gravel path, but also in the house:

Partner: you always have to aim for the middle of a tile

Alan: Ah, yes

DB: But that was then seemingly also with a crutch, or not?

Alan: Yeah, when I had that crutch here for the first time

Partner: here in the house, it was dreadful

Alan: Exactly, had I told you that?

DB: no

Alan: That tip [of the crutch] had to be placed in the centre of the tiles, but also if I walk here, then I have to do it too, on those tiles, but on the market square with all those little clinkers... *shakes his head* and now I don't do that very much anymore, but I also don't walk in the house with that crutch anymore. Here [in the kitchen] I have more movements, changes

DB: Movements? Changes?

Alan: That you sit down, or walk, or stand somewhere, or do the dishes.

From Alan's statement it seems that objects, or better, the collection of object dimensions have certain thresholds in size and amount that evokes particular compulsions; faced with thousands of small clinkers around one's body extended by a crutch, it would be an insurmountable compulsive task. As size and amount are relative to perception, a spatiality emerges in which a landscape of tension can erupt into compulsions, and outside of which such 'danger' does not necessarily exist as such. This latter example also confirms that materially and socially bounded spaces that invite certain activities, such as doing the dishes in the kitchen, watching TV in the living room, and walking to wait for the train on a railway platform, also produces the possibilities for bodies to become involved in certain compulsions.

In short, compulsive configurations emerge as spatial constructs of subjectivity that are made by the spatial formations of the body, self-referential objects, objects-in-excess, and collections of such excess when they coincide with in simultaneous movement. There thus seems to be a pattern or rhythm to such simultaneous movement that creates and recreates a compulsive realm. The next chapter explores such realms, and how the ecologies that they constitute create a sense of well-being, and how it is through such ecologies that well-being in place is negotiated or even forged.

Notes

1 These shifts could often be felt by myself as a spectator.
2 Objectness here is derived from Jane Bennett's construction of affective objects that she notes as becoming powerful in the social world through their vibrancy that emerges in material relations of human–non-human ecologies.
3 Selective serotonin reuptake inhibitors (SSRIs) are used as antidepressants in treatment of depression and anxiety disorders.
4 Individuality here refers to categories of significance, e.g. 'this is a vase, that is a phone', that's a window frame.
5 The original Dutch word she uses is 'prikkel' which does not exist precisely in English, but is a collection of the more formal 'stimuli' and more informal 'poke', 'itch', and '(pin)prick'.

References

Anderson, B. and Wylie, J. 2009. On geography and materiality. *Environment and Planning A* 41, pp. 318–335.

Ash, J. and Simpson, P. 2016. Geography and post-phenomenology. *Progress in Human Geography* 40(1), pp. 48–66.

Bliss, J. 1980. Sensory experiences of Gilles de la Tourette syndrome. *Archives of General Psychiatry* 37, pp. 1343–1347.

Davidson, J. and Henderson, V.L. 2010. Travel in parallel with us for a while': Sensory geographies of autism. *The Canadian Geographer/Le Geographe Canadien* 54(4), pp. 462–475.

Hoel, A.D. and Carusi, A. 2015. Thinking technology with Merleau-Ponty. In: Verbeek, P.-P. and Rosenberger, R. eds. *Postphenomenologial Investigations: Essays on Human-Technology Relations*. London: Lexington Books, pp. 39–56.

Leder, D. 1990. *The Absent Body*. Chicago, IL: University of Chicago Press.

Lingis, A. 1998. *The Imperative*. Bloomington, IN: Indiana University Press.

Lingis, A. 2000. *Dangerous Emotions*. Berkeley, CA: University of California Press.

Manning, E. and Massumi, B. 2014. *Thought in the Act: Passages in the Ecology of Experience*. Minnesota: University of Minnesota Press.

Meige, H. and Feindel, E. (1907 [1902]) *Tics and their Treatment*, with a preface by Professor Brissaud, trans & ed Wilson, S.A.K., New York, NY: William Wood and Co.

Merleau-Ponty, M. 2013 [1945]. *The Phenomenology of Perception*, trans. Smith, C., London & New York, NY: Routledge & Paul Kegan.

Seigworth, G.J. and Gregg, M. 2010. An inventory of shimmers. In: Gregg, M. and Seigworth, G.J. eds. *The Affect Theory Reader*. London: Duke University Press, pp. 1–28.

Simondon, G. 2005. L'Individuation à la lumière des notions de forme et d'information. Editions Jérôme Millon, coll. Krisis. In: Manning, E. and Massumi, B. 2014. *Thought in the Act: Passages in the Ecology of Experience*. Minneapolis, MN: University of Minnesota Press.

Souriau, E. 2009. *Les différents modes d'existence*. Edited by Isabelle Stengers and Bruno Latour. Paris: PUF. In: Manning, E. and Massumi, B. 2014. *Thought in the Act: Passages in the Ecology of Experience*. Minneapolis, MN: University of Minnesota Press.

Williams, D.W. 2005. *Autism: An Inside-Out Approach: An Innovative look at the Mechanics of Autism and its Developmental Cousins*. London and Philadelphia, PA: Jessica Kingsley Publishers.

Wylie, J. 2006. Depths and folds: On landscape and the gazing subject. *Environment and Planning D: Society and Space* 24, pp. 519–535.

7 Compulsive durations
Ecologies of stability

In suspension of other-than-compulsive life, compulsions are performed when the body and elements of its surroundings tune into simultaneous movement and configure towards and away from each other. In contrast to the neuropsychiatric and societal position of the human as a 'perpetrator' of compulsions, compulsivity may be better understood as a process that captures the movement *of* the body, not *by* it. To different extents and in different ways, bodily surroundings effectively sweep people off their feet when they become enlaced in a compulsive situation.

This chapter inflects the spatialities with which compulsions emerge as located in conjunction with everyday spaces. It explores how these spatialities produce 'maps' of body-world formations, and how that leads to a re-conception of *body-space* as emergent from, and ordered by, compulsivity. To this end, it traces how this body-space, consistent of compulsive systems becomes negotiated: in particular, the effects of the accomplishments of compulsive interactions when other-than-compulsive life resumes. The just-rightness of reconfigured bodies and objects outlast their configurations, as glasses stand in perfect rows in the cupboard, knife tips have been defined in their sharpness and fingertips in the soft fleshiness, and small items have been grouped on a desk. In the legacies of their emplaced performances, compulsivity seems to have a brighter side than the compulsions seem to have.

Situations: Systems and space formation

Not all objects reconfigure towards a compulsive situation in this way; some retain their vitality through persistence beyond one situation. The existence of such objects does not give rise to an 'anti-geographical' or decontextualised argument, on the contrary, it demonstrates compulsive non-human vitality in a stable multiplicity of appearance. As such, their vitality is so strong that they can reconfigure many situations towards compulsive engagement. The persistence rooted in a sort of outspoken individuality is then composed of excess collections that only result in compulsive movements with the human body, and no other kinds of bodies or objects. For instance, Sage's partner is an amateur sculpture artist, and through his experience of living with

DOI: 10.4324/9781003109921-8

her for a few years prior to the study, he had observed her compulsively engaging with her surroundings. With this knowledge of sensibilities that pervade interactions, he made a marble statue for her that protrudes in several peaks, collapses into itself, and forms curving ridges that direct the gaze. She is astounded by its just-right perfection. When she holds it, she needs to press her upper lip and nose onto it and tap it lightly with her teeth and tongue. She glides her fingers along the more dramatic sharper edges, pushes her thumb into a dip, pulls it along a valley, and pushes her index finger into sharp and blunt tips. The collective coldness, slickness, curviness, and edginess that articulate in the stone statue configure Sage towards the intimate compulsive situation:

> When I just had it, I would hold it and interact with it very often. Not so much now anymore ... but now I have it in my hands again, I think 'ooh, that is really very nice!' *laughs* So good! *laughs* but if it would be in front of me, I'd often hold it for a bit. Now I feel... now I want it! *we laugh when she picks up the statue standing before her*.

This statue was one of the few objects in this study that always already exists in a compulsive world and only incites bodily movements that are compulsive. Incessantly extending an invitation for compulsive enlacing, it retains its affectivity over the years and months it had been in Sage's life. From all angles, at all times, through all situational changes, this object remains welcoming to compulsive configuration of the body, as such rendering it a powerful object. Its presence in the room it is kept in intensifies the landscape of tension and continually elevates the space as a distinctly compulsive world.

Even when situations of bodies and objects remain relatively static, such as whilst sitting on the sofa and using one's phone, reconfigurations towards compulsive situations take place. Situations that are more dynamic can be experienced as violent in their equally dynamic configurational demand with the body moving with collections of object excess that rapidly recompose in formations of multiple patterns, lines to be looked at, felt, heard, and smelt. For this reason, Lowri struggles with cycling and Ginny cannot always enter supermarket when she has had a busy day:

> I haven't cycled in a long time, which was like, I was so overstimulated. It just didn't work; cycling in the city; 'hold up, and that one, and that one, and that one'. So busy! 'I have to watch out for that, and I have to stop there, and there's a tram and there's a bus'.
>
> (Lowri)

> For some reason I just had a blockage, and that ... Yeah ... I couldn't even reason through it, it was as if I was stuck or something, as if the brain I needed for grocery shopping had just failed.
>
> (Ginny)

The abundance of elements that demand Lowri's body to reconfigure accordingly keeps her occupied, not allowing her to configure towards cycling to the extent she needs to keep herself safe in the traffic of the city centre where she lives. Similarly, Ginny has been in supermarkets where she "froze" and could not sufficiently concentrate on buying groceries. The multiple invitations to move with the intricate, rapidly emerging and vanishing patterns of colours, textures, lines, temperatures, and shapes, weights, and 'shine' "blocked" any other-than-compulsive, meaningful and purposive perception and movement, even when standing still. In the multiplicity and incessant dynamism that can capture a body, a body is rendered still. Stillness does not equate with passivity, but the impossibility to tend to opposing claims of movement that emerge simultaneously; there is simply no time to compulsively engage with one patterned collection of objects-in-excess because another one that may be composed of the same objects also need tending to. One can think of it as the spectacle of multiple fireworks exploding closely together and overlapping in their radius, in Lowri's case, or in Ginny's case the whole show at once.

Whilst Sage's statue's incessant self-organising compulsive vitality retains her configuration in a compulsive enlacing beyond any situational change, other objects acquire a vitality exactly in their distinction from the material context in which it emerges. 'Manmade' objects, as Alan puts it, and deliberate situations of objects that pervade a presence of humans in 'natural' areas can configure his body to immediately compulsively seeing[1] them, changing his path, and feeling agitated. In the wooded area near his home, wooden fence poles stand out for their out-of-placeness (after Cresswell 1996) in the woodland, which often compulsively configures him towards them, causing him difficulties configuring back to walking. "Nature is allowed to be messy" according to him, and the straightness and long lines of the poles stand out. Although Alan puts the distinctive dimension on a nature–humankind divide, despite the poles being made of natural wood, his allusion to messiness may be more productive to understand the compulsive tensions. Indeed, as straight lines in nature are rare, and in plants even more so, the compulsive tension of the situation may more likely emerge from the rationalised forms that disrupt the other-than-compulsive irrational forms, with messiness equalling irrational forms that cannot be ordered. Thus, by virtue of the relatively similar 'irrationality' of the surroundings, and an equally calm landscape of tension, this kind of compulsiveness can occur more easily because in other situations a wooden pole would not have stood out to such an extent that it needed to be compulsively interacted with.

A similar kind of compulsive tension often surrounded breadcrumbs on tables, tending to be removed. Despite their small size, they stand out on otherwise smooth, shiny, and uninterrupted surfaces. Both Alan and Siôn kept an eye on crumbs appearing on the table and immediately wiped them off the table to gather them in their hand or pick them up individually and drop them onto a napkin or into a breadbasket, which kept both men continually occupied during lunch; especially Siôn, who needs to closely watch his children when they enthusiastically eat their sandwiches for lunch. For Alan, this vigilance can continue until after all plates and cutlery are removed from the table, and only after having wiped the whole table with his hand can he fully configure back to other-than-compulsive life. For both men, the surfaces of tables simply cannot be disrupted by crumbs. For Siôn, this compulsive forcefulness also emerged from spotting white fluff on soft dark blue carpet or on the camel-coloured stretched cotton of a sofa.

What is held in common between specially designed statues holding a grip on rooms, wooden fence poles disrupting 'nature', crumbs interfering with the slickness of table surfaces is a mutual, relational compulsive constitution between objects that do not involve the human: the way it is perceived does not play a role, nor does the kind of interaction. Rather, the human becomes involved *only to correct the situation* of objects in accordance with each other. Hence, objects – and other bodies – do not only reconfigure into compulsive enlacing with the human, they also retain a compulsive configuration between themselves.

We have already established that (not-)just-rightness can be entirely situated in the corporeal constituencies and exist independently from human involvement, notwithstanding that in the absence of the human the compulsive tension does not exist. This can, however, be extended. Indeed, such 'corrective' compulsive involvement of the human can be thought to create, maintain, or break 'systems' with which compulsive configurations take place. Such systems consist of all the material entities that are needed to make a situation just-right again, including the multiple overlapping categories of objects, human and non-human bodies, and elements that make up a space[2]. For instance, Elisa's compulsive ordering of the vases on the windowsill can be regarded to take place in accordance with a compulsive system in the living room. A vase with plastic yellow tulips on Elisa's windowsill is configured in accordance with a similar vase with plastic yellow tulips, with the depth of the windowsill itself and with the books on the windowsill, but not in accordance with the other tulips in the living room, nor with the curtains, other similarly coloured entities in the room, or the woodgrain of the window frame.

Compulsive ordering is then better understood as completing, supporting, or stabilising compulsive extracorporeal systems to which objects and human and non-human bodies can 'belong' because of their shared appearance dimensions. Following a Deleuzo-Guattarian vitalism, these bodies are assembled in their desirous production of just-rightness. Just-rightness is, however, not a function of the assemblage itself, but precisely only *the way in which* human and non-human entities that form compulsive systems become explosive and start to configure people into compulsive interactions. Objects and bodies can also 'travel' between systems when their situation changes or appearance shifts, and they 'meet' or 'coincide with' bodies' excess that is part of other systems, objects, and bodies. A situation that sees objects travel between different systems is for instance Ginny's window-sill. The different compulsive orderings of the plant pots, figurine, and other elements are rooted in multiple co-existing systems that gain precedence on the basis of the "kind of itch" she gets and from what side of the coffee table she approaches it.

Thinking through compulsive systems opens up a productive way to understand and further explore how compulsions can repeat themselves and how some are so repetitive that they are anticipated to be performed and accommodated for in other-than-compulsive life. Feeling the same sensations and doing the same movements in the same place, suggests that sensation, action, and material organisation are bound-up together. As such, it indicates some sort of spatialisation of the processes of perception. Therefore, we turn towards Deleuze and Guattari's conceptualisation of the *percept*[3], to integrate appearance and object and body excess with their spatial situation. On the basis of Bergson's object images (1911), Deleuze and Guattari develop percepts as unique constellations of all that is present: gatherings of the subject, object and perception itself, which renders them "independent of a state of those who experience them" (Deleuze and Guattari 1994 [1991]: 164). As "anterior to the prescription of perceiving subject and perceived world" (Anderson and Wylie 2009: 332), percepts constitute "a pure flow of life and perception" (Colebrook 2002: 74) and acquire quality by the capabilities and sensibilities of both object and subject (Dewsbury et al. 2002). Percepts can then be thought to float or hinge in a situation of various materialities and perceiving human bodies. It follows that collections of excess spatialise *with* and *as* percepts, and the percept can be regarded as a spatialisation of compulsive *movement with*. Indeed, without necessarily predetermining compulsive engagement, when percepts are 'stepped into' it is felt as the start of the compulsive configurational process and when not disrupted in this process. For instance by speeding up and not registering certain sights as Ginny contends, comprehends and reproduces a just-right situation. The percept is thus the individuation process of a compulsion (after Simondon 2020 [1964]), and the compulsive situation can be understood as 'crystallisations' of just-right compositions. After compulsive interactions have finished, just-rightness peaks, and the volatile energy landscape dies down a little. Compulsive percepts retain their potential through constant (re)configurations, as each enlacing and actualisation of system configuration leaves a historical mark, that is, the experiential repetition of becoming enlaced in a compulsive system. The ways in which the need to perform compulsive interactions can be negotiated to a certain extent can also be understood better by utilising percepts and their collective existence and their spatial distribution.

What we end up with is thus a way to understand the spatial and sensory relationships between perceptive processes and systems through the just-right principle. How compulsive systems as organised percepts can configure the human body towards compulsive interactions and play a role in ongoing life can be considered using Erin Manning and Brian Massumi's (2014: 9–10) conception of the entanglement of movement and perception:

> Your perception is focused on the coming and going of the openings, which correspond to no thing in particular. Each opening is a field effect. It is an artifact of the moving configuration of the bodies around you, factoring in their relative speeds, and their rates of acceleration and deceleration as their paths weave around each other and around obstacles. The opening is not simply a hole, a lack of something occupying it. It is a positive expression of how everything in the field, moving and still, integrally relates at that instant. It is the appearance of the field's relationality, from a particular angle. The particular angle is that of your body getting ahead. The opening is how the field appears as an affordance for your getting-ahead. Your movement has to be present to the opening as it happens. (...) You have performed an integral dance of attention, seemingly without thinking. But you were thinking, with your movement. Your every movement was a performed analysis of the field's composition from the angle of its affordance for getting-ahead. Entering the dance of attention, your perceiving converged with your moving activity, and your activity was your thinking. (...) But in the mode of environmental awareness that effectively got you to the office on time, it was not the object "sidewalk" that afforded the last leg of your commute. It was the fleeting openings, now forgotten. The openings are long gone. The sidewalk remains. The stability of the sidewalk, its ability to re-feature in experience from moment to moment, is an enabling condition for the ephemerality of the openings.
>
> (Manning & Massumi, *Thought in the Act: Passages in the Ecology of Experience*, 2014, p. 9-10)

Manning and Massumi thus create a world of continual flow that is governed by perceptual openings, pulling people forward and giving rise to vital constellations of possibility. In addition to other-than-compulsive openings that are produced by and within a space, compulsivity marks another set of mostly unwanted openings that need to be tended to, which effectively close down other other-than-compulsive openings. Compulsivity in its relation to non-compulsivity then becomes visible as *additional layer of liveliness* betwixt and between other-than-compulsive formations of liveliness, as compulsivity can be recognised to open up new and unknowable possibilities to interact with the corporeal surroundings. Amongst other things, understanding compulsivity in these terms explains Lowri's inability to cycle, Elisa's struggles with driving (Chapter 3), and Dylan's difficulties with finishing vacuuming his room (Chapter 5), but also why Siôn enjoys looking at the just-right home office he constructed (Chapter 5).

As argued by Manning and Massumi (ibid.), opening and closing worldings entails all perceptive processes. In the context of compulsivity, this distinctly includes touching, pushing, feeling, biting, and other tactile interactions. Similar to orderings, these kinds of compulsions amount to a collection of spatially dispersed demarcations of *movement with* different articulations of the materiality and fleshiness of bodies and embodied objects. As collection, the intensity and violence of these demarcations become especially apparent in spaces that hold many particularly evocative objects. Mina narrates a recent experience in a museum:

> Clothing from tsars, it was astonishingly beautiful, but they were placed behind glass, and then I can … ehm … it's like being given a plate of tasty French fries and you're not allowed to take one, and that made me... I really wanted to touch them really badly, and also those items that weren't placed behind glass, in no way I could get to them.

The organisation of the pieces of clothing, the way they are lit, their glass displays, and placement out of bodily reach give rise to a certain highly evocative character. Museums and other spaces then actualise with the human moving through as a particular affective regime. This constitutes spaces as entailing and individuating as distinct 'compulsive cultures'. These more-than-human cultures emerge from a particular organisation of particular compulsive systems that cut through, intersect, order, and shape other-than-compulsive worlds. In effect, spaces become gatherings of systems of human and non-human configurations that always have the potential to explode into compulsive enlacings via interactions between the corporeal and the extracorporeal. Certain compulsive cultures that envelop spaces are more conducive to other-than-compulsive life than others.

Futuring compulsions: Stabilisations

As established, just-rightness is the most powerful moment of compulsive interaction and marks the optimal configuration between bodies and objects. Indeed, acting compulsively to make a situation just-right can bring bodily environments into a 'state' in which the emergence of future compulsions is omitted, or at least postponed. As such, it indicates a particular, potentially conducive, spatial order of bodies, objects, and spaces, and therein constitutes a spatial ordering principle. This order, as constituted by the situated, floating percepts provides further possibilities to explore the legacies of individual compulsions, their combined legacies in a space, as well as possibilities for mediating them. This section establishes what it means when compulsions have instated a just-rightness; is it order *as such* that qualifies the just-rightness? Or are there different kinds of 'accomplishments' that figure in people's further live after their body has reconfigured after the compulsion?

In several interactions that involve touching, its legacy in the form of its accomplishment seems to have a confirmative aspect. Echoing experiences of many other participants, during a walk, Joe needs to compulsively touch a chestnut upon laying eyes on it and stare at it:

> In that first instance you think 'it's a chestnut', but before you're convinced, you need to pick it up, and then you touch it, and then you need to ensure if it really is a chestnut by looking at it.

The 'chestnutness' of the object needs to be established three times in this instance. It seems that only after the final interaction it acquires a just-right sense affirmation of the textures, shine, colours, and volume that constitute this object as chestnut to Joe. The chestnut-object falls apart into these different dimensions and form a disintegrated compulsive system on the ground. Only upon separating it from its context of the ground, touching it, and staring at it long enough it reverts into a chestnut and can he continue walking. As Sage also contends when pressing her hand flat onto the kitchen table: "it's a fraction, you know, that I hold the material just that little bit longer". Such 'confirmative' compulsivity can be thought to reinstate objects and other extracorporeal entities through intense inspection and/or by virtue of altering its context. What we see is that although compulsivity might be governed by perception and movement, following the accomplishments of such 'confirmatory' compulsions points towards a need to ontologically de-centre the human body.

A further compulsive interaction that is testament to unsettlement of the human category in compulsivity takes place between Ginny and a shopping cart when she leaves a supermarket with a full load of shopping. When she pushes it towards the car, the cart seems to require retaining the direction it faces when entering the parking lot. Repeating this several times, Ginny walks around the cart to take out the grocery bags, holds them up in the air whilst she walks back around the cart to manoeuvre herself in the doorway of the car where she can put the bags on the backseat. During the effort, the cart does not move, despite its wheels and lightweight. When she is finished putting the bags in the car, whilst not changing the direction it faces, she pushes it backward and sideways to the parking lot stall. In this series of functional acts, or intricate, partial, but singular compulsive interaction, the cart takes the central point of the situation, and Ginny('s body) moves around it.

This requires an ontological re-centring of compulsivity to the cart in person-place relations that radically diverts from clinical conceptions of compulsivity, as well as from humanist geographical traditions that centre the human body in embodied action. Centring non-human objects in compulsive interactions explains many more types of compulsions. In particular those that accomplish a securitisation of a number of non-humans in 'their place'. As such compulsions play an important part in her everyday life, Sara expands:

> I find that that chair belongs there, I just really like it when things have their own place. (...) The other day, I'd been looking for my bicycle keys

for almost one and a half hour, and I had put them right there on that thing *points to a chest of drawers*, but then I am completely lost where in the world they could possibly be, because they're always there." *points elsewhere* (...) shoes are separate things, they don't belong together.

She explains that compulsive interactions need to be performed when her home is altered in one way or another, for instance, when objects move, it can leave her 'completely lost'. This is particularly clear when she cleans her home, as the combined efforts are an elongated compulsion or a compulsive endeavour. After cleaning a windowsill with a wooden figurine on it and moving on to other parts of her home, and after walking past it a couple of times looking at it, she repositions it slightly. Cleaning her entire home becomes a way to ensure that all objects are in their 'rightful' place, which she does every morning for about 35 minutes before she "is allowed [to have] coffee". Although she ends up with a clean house, cleanliness is not what drives the effort: she only moves on to the next (part of a) room (e.g. windowsill, kitchen counter, work station, bathroom, etc.) after having interacted with different objects in particular ways. For instance, she picks up a tiny figurine for wiping the windowsill underneath it, putting it back, and stroking it with her fingertip. Thereafter, she very carefully slides the cleaning cloth over the entire surface of the windowsill but carefully going around the figurine: not touching it as to not reposition it. Further on, when she has finished using cleaning cloths, she picks up the cleaning cloth holder sleeve, opens and closes it seven times whilst standing in the middle of the room before walking down the stairs. These compulsions all help to ensure that the holder sleeve is definitely closed, the carpet definitely does not have crumbs on it, and the figurine definitely sits on a dry surface that is otherwise definitely free from dust and crumbs.

Sara's body becomes the measure through which situations and the elements that make up such a situation – including both the body and the objects – are confirmed in their place. This confirmatory compulsive effort is only successful when having perceived the just-rightness of the organisation from specific angles and through specific bodily motions. Sara's compulsivity resonates with Tomos' who asserts: "if I have to oversee the present and the future at the same time I become very anxious". The confirmative aspects of compulsive interaction that crystallises the situation into the just-right state is an indication of a *stabilisation of the situation* after the compulsive aspect dissolves. By extension and in turn, compulsive interactions are better understood as addressing an *instability* or a *breach of a previous stability* in a space. The latter emerges from, for instance, Elisa's house which consists of particular stabilities in the sense that some objects are not to be moved. She asserts that the organisation of bookcases on the living room wall cannot be changed, and only new books can be added if they resemble the ones already on the shelf. This fact is so strong that the fit of a book on the shelf is already taken into account when buying a new book.

She elaborates on the consequences of buying a book that she wants to read but does not fit:

> Elisa: For instance, Jo Nesbø has one book, and I took it to the charity shop because it's bigger than the others (...)
>
> DB: Have you read it?
>
> Elisa: Yes, and then off with it. I couldn't have that in my house.
>
> DB: So that it never lived anywhere?
>
> Elisa: Well, very briefly underneath [a stack on the windowsill] *points* and it irritated me to such an extent that I've thrown it out.

The composition of the bookshelves on wall can be understood as having stabilised and is actively maintained by virtue of careful compulsive selection of additions and organisation of new books and other items. This implies that the compulsive interactions that had stabilised this situation indeed prevent the performance of new compulsions, and, as such, refrain other-than-compulsive life from compulsive anxieties and disruption.

Elisa's compulsive stabilisation of her bookshelves has a long duration in comparison to a more fleeting stabilising compulsions performed by Cai. He aligns his shoes with his laundry basket that he had just put on the floor in the middle of the room, after having put his folded clean laundry in it. After having attended to other objects and having done other chores he picks up the basket, thereby breaking its stabilisation with the shoes. Hence, compulsive stabilisation acts can produce dynamic compositions that are more or less fleeting. Nonetheless, such interactions can result in stabile compositions of relatively dynamic objects, and the mobility of such compositions hinges on the ease with which these objects can be moved (i.e. a large item is more difficult to move than shoes), and the ease with which objects configure into new compulsive situations.

When taken out of a stabilised configuration, bodies, objects, and spaces – by virtue of e.g. corners, plinths, wallpaper patterns, flooring textures, colour combinations – can be experienced to have altered configurations that may change the dynamic of the old configuration. Whilst cleaning the bookcase in her living room, the objects Sage picks up, cleans, and places back often do not end up in the same composition as they had been in before, because they came to appear differently and therefore configure differently amongst other objects. Moreover, for her and others, the human body can be actively part of the stability itself beyond the interaction. For example, when Sage puts clean dishes from the dishwasher into the cupboards her right thumb, pinky, or index finger were involved in rubbing and/or pressing glasses, cups, mugs, larger cutlery, spatulas, and other large items before she put them in their final place. She expands:

> I'm usually holding items in a specific way. If it's put in its place, it's usually alright, but it's especially the process of putting down that needs to happen in a very particular way.

In this compulsion, the body, its movements and its weight, tensed muscles, and stretched skin hold a just-right stability in conjunction with the clean, still warm object, its trajectory from the dishwasher ending in the cupboard by its pressure on the hard slick surface of the shelf next to similar objects. This specific interaction can still be considered to constitute a stabilisation, despite the compulsive constellation being mobile and fleeting. Another such mobile stabilisation for Sage involves another human body:

> For example, often I like it better when people walk on my right side, rather than my left side. That has to do with a feeling of balance, I think, like 'right is better'. (...) I'm also finding it very difficult, for example, to walk on someone's right side, so I want them to be right of me, and me on their left side.

Here, the just-rightness of the compulsion seeks to achieve, manoeuvring towards, and remaining on the left side of the person, and the stabilisation locates in the positioning of Sage's body in relation to another body, regardless of them potentially stopping, gesturing, running, etc.

To put this argument in the context of the systems that produce compulsive cultures, stabilisations are ways to order and diminish the effects of the diverse systems at work in a space. A system that is 'strong' and stabile retains its vital coherent configuration from multiple angles. For instance, Elisa constructed a spare bedroom[4] from scratch, which is "entirely just-right", that is, unless someone sleeps there, after which she brings the space back to its just-right state. She also decorated a living room wall with books and other items in such a way that everything "has its own exact spot". A weaker system might be easier to break in terms of similarity in appearance and mobility of objects and bodies. For instance, desks, dinner tables, and kitchen counters, and Bill's carefully organised fridge, which is also frequently 'interfered with' by Bill's partner and son. Such systems can require constant (re)configuration through compulsive interactions towards sustainable situations.

As compulsive interactions can stabilise the appearance of an object or a bodily movement one is involved in, this produces an enlacing of the body with its constituencies in compulsive interactions that constantly require stabilisation or demand a renewal at different speeds and with different trajectories. In turn, these stabilities influence the emergence of future compulsions. This implies that compulsivity as a medicalised phenomenon that, to date, has been understood as *exclusively human*, can thus be reconceived as ordering principle of the human in the spatial relations between the body and the extracorporeal. 'Distributing' the condition beyond the body then renders these medicalised performances truly more-than-human and reconstitutes them as a *more-than-human condition* (Beljaars 2020).

The conception of compulsive interactions on the basis of movement and perception of which their accomplishments can be conceived of as situated and emergent stabilisations has several implications for body geographies. These stabilisations can take the form of object compositions that exclude the human body and are compulsively interacted with to retain its stability outliving the interaction. Rooted in object-oriented ontologies (e.g. Harman 2002, 2018) and technoscientific approaches to objects (e.g. Ihde 2009), this is in line with Ash's (2015) post-phenomenological analysis of tinnitus, which is produced as a result of the indirect and unintentional relations between the inorganically organised speaker and sound system objects. Following Ash's demonstration of experiencing the ongoing beep sound long after having been removed from the situation and enduring affective object constellation, the next section traces the spatiality of compulsive stabilisations as enduring.

Ecologies of stabilities

In the knowledge that compulsive interactions produce and reproduce stabilisations of collections of objects and human bodies, this section explores how to reconceive spaces through the ecologies incited and sustained by compulsive interactions and the enduring affects they produce. Compulsivity is not only mobilised as *mediator* of space, but also as *active producer* of space through the stabilisation efforts of the systems that pervade them. For instance, Cai can be seen to fold back a corner of bedlinen, which has a particular stability in the compulsive interaction involving the bed, the linen, his position of being outside the bed, and noticing it in addition to other elements. Another example is that every time another person touches the big cushion on the chair in the living room, Siôn walks over to puff it up. This requires particular human presences and absences in that room – amongst other things – the chair and the cushion, daylight, and him walking around and stepping into different percepts that incite body and object configurations. With different durations of present or potential stabilities, particular cycles of stabilities break up in a space. For instance, Alan picks up three objects from the garden floor and puts them on the garden table, as he and his partner had sat outside enjoying the summer warmth the evening before. This is the only stabilisation the garden needs because the plants in the garden do not require any compulsive interaction, because they are 'natural'. As such, people come to know and create spaces through learning and adapting the *patchwork of stabilisations*.

Homes and other private spaces are environments that are very well known, which seems to have two main consequences. Firstly, having elaborate knowledge of the details that make them up seems to produce intricate and complex stabilisation requirements. Indeed, familiarity with a situation, place, situated object speeds up the configurational process, because the preceding times that the compulsive act has taken place has created a sort of memory, or a 'passive registering', of such an act. With their increasing in-depth knowledge of spaces, people tend to perform more recurring

compulsions. Paradoxically, as there are more possibilities to incite configurations familiarisation with places, as well as very busy environments with
much fine details can then add to the exhaustiveness of residing there. This
effect may be mediated by feeling more relaxed at home for some people.
Residing in a different place from the home for a while also takes away the
requirement for a while. Sage argues that going on a holiday also means
"going on a holiday from [their] Tourette's!":

> When I went on a holiday, a week with my parents, and because all my
> compulsive acts are so related to places, objects, furniture, the moment
> I was on a holiday I had lost it all, because all those things, like the
> route from the living room to the kitchen. These things that I have to
> touch ... those were different there, and probably if I'd been there for
> three months, I would have developed new ones at some point, but in
> a place where you're not residing for a long time you have less time to
> eh ... you know.
>
> (Sage)

> Those are all things I own, I see that, I see everything as an extension of
> myself (...) when I go to my mum for the weekend I decouple from here
> and completely recharge, because at hers I don't feel the compulsivity
> and I can't do anything, you understand? Then I'm in another house,
> because it's someone else's.
>
> (Ginny)

Alan with strong sensibilities for straight lines and the lack thereof, experiences the garden shed of his neighbour that he can see from a first-floor
window, to not adhere to a sense of just-rightness:

> I look at it every day. Every time I'm upstairs, but it doesn't irritate me
> that it's so strange: that it's not quite right.

Both he and Ginny resonate that a lack of specific 'ownership' or responsibility over situations that could have enlaced them in compulsive interactions. Indeed, Bill only pinches, and presses, items in his shopping basket
that he intends to buy, and he does not compulsively alter the appearance of
shop furniture or items he does not buy. Nonetheless, this does not say that
spaces he has no ownership over do not require compulsive stabilisations.
On the contrary, stabilisation requirements articulate a different kind,
which might, for instance, include more touching or bodily positioning
interactions, for instance, stepping over painted lines or those that protrude
spaces from corners or furniture.

Secondly, a deep knowledge of a place such as the home also provides
many possibilities to stabilise, as the participants have high levels of authority over these spaces. Siôn reminisces over the flat he lived in before moving

in with his wife. Because he was completely in charge of the organisation and the only cause for change in his house, he called this living situation "planet Siôn", and narrates it as a utopian place:

> Then I would cook, that I needed pans and things of course, so that changed [the organisation of the kitchen], but other than that it remained always exactly like I wanted. Then it was perfect, so when I'd come home, terrific! The remote control was still there, the cushion stood like that, always neat and tidy, really very orderly. Yeah, just great.

Despite thinking about living under these conditions as preferable, he also thinks of residing in such places as problematic and therefore unattainable: "It's just like taking to your planet again. There we also find things that aren't right, it is exhausting, you know". Also Elisa's living room has been arranged and decorated by her with strong ideas about the appearance, location, and potential for mobility of each element, except one small table which is her partner's 'patch'. When she cleans the living room, she explains how elements are connected to other ones and for what reason, such as the aforementioned tulips and wall with bookcases. When she arrives at her partner's patch, she refrains from touching it, despite the frustrations it conjures up. Upstairs she has a room that contains wood she dries: she asserts that "it was allowed there, because the disorder has a function", and "it was an intermediate solution". In effect, the room had become a parameter that was beyond stabilisation, hence she does not like going in there.

With the collections of stabilities people create parameters of relative intensity that are produced by often having to reinstate stabilisations. In addition to these parameters, spaces come to exist of parameters of relative peace that are produced by sustainable stabilities, ownership, and/or the absence of other people, and parameters that accommodate for situations expected to change. An example of the latter parameter is the collection of places the remote control can be expected to locate in Bill's living room. Much like Siôn, he is not fond of the remote control changing location in general, but as it is an often used and highly mobile object, he is happy to find it in three other distinctly designated locations. These designated locations are limited to the edge of the large cabinet, the armrest of any living room chair, and the coffee table. The remote control can thus 'roam' these locations, but cannot leave it, or its absence produces a compulsive situation in which Bill is required to find the remote control and compulsively place it on the right-arm rest of his chair. Therefore, being able to create and negotiate the different parameters that produce the appearance of spaces allows people to live their lives as little interrupted as possible.

The sizes of such parameters seemed to differ, with Elisa's living room, kitchen, and back garden seemingly capturing relatively small parameters. The spaces bursting with carefully arranged small and colourful patches lead her to assert that "it seems all a bit messy". Considering larger patterns,

Tomos sees the appearance of his flat in accordance with larger colour surfaces, lines, and textures, such as emergent from combined excesses of curtains, sofa, large chandelier, and tall plants. The smallest stabilised parameter in the spaces of his downstairs flat involves a bulletin board, which he pleasantly describes as "just nice and casual" and "as unorganised chaos". Also Mina's living room has distinct parameters that she stabilises as much as possible, which seems to alter with the activities she chooses to do. For instance, she likes sitting on the sofa to watch TV, whilst when sat on the sofa to read, she requires a particular body posture that allows her to do so:

> If I'm, for example, sat here... this lamp irritates me so much when I read my book *points* (...) so I have to get rid of that, then I will now (inaudible) because of that blue light *points* – incredibly irritating, so that will be turned off. (...) It's turned off now, but if that red light is on ... *points* that's impossible, that needs to be turned off as well. That little lamp also needs to be off. I can't read then either whilst needing to look at the page, that little lamp is just being irritating. 'Nonsense, just read!' but then I'll just eh ... usually I'll sit in that sofa corner there *walks over to the sofa and assumes her reading posture* and that's not even that uncomfortable.

Mina's living room is very softly lit, which gives all presences a greyish/cream-like hue. Colours such as "those yellows [post-it notes], they give a kind of irritation". The only other colours can be found in the painting on the far wall across the table from the chair she usually sat at. At times "that also irritates [her]". She points out that particular patches of red were too bright for her. As the painting does not hang at the exact height the other paintings in the room, it hinges on the threshold of non-conformity with other paintings with which it requires stabilising. As such, the painting that is relatively rigid in its appearance on that wall can hardly be stabilised, whilst the small bright lights can be turned off when necessary. Also, the charcoal drawing she makes during an eye-tracking session has to be removed from her house, "otherwise I'll have to continue with it (...) Then I don't want to see it, because I'm finding it *not right*, and then it needs to go". Individual objects, such as the drawing, can thus be so powerful that they seem able to overthrow all careful stabilisation parameters that produce the space (also see Chapter 6).

Indeed, individual presences seem to be able to work as a catalyst or as a neutraliser of destabilisation throughout spaces. Apart from strongly affective – and therein powerful – objects, other human beings can become such presences. Sage recalled her mother becoming such presence when her mother feared compulsive interactions would harm Sage. For instance, in relation to a hot oven dish, Sage would have configured herself in accordance with the hot oven dish in such a way that she could cope with the urge to compulsively interact with it until it would no longer cause skin burns upon touching. Her mother would then undo such configurational 'precautions' by saying "watch out [Sage], that's hot". However, echoing pretty much all people with Tourette sensibilities who argue that well-intentioned warnings actually amplify the intrigue, Sage states:

> By saying that it is hot, it just becomes a massive trigger again, leading me to keep having to think [Expression of excitement] I want to touch it! *laughs*

Following on from the conceptualisation of compulsive interaction accomplishing stabilisation, other humans are perhaps very difficult – if not impossible – to stabilise in combination with the space resided in. The presence of people might unsettle earlier stabilisations due to changing bodily positions and one's ability to, and the social appropriateness that allows for altering and repositioning non-human objects. Ginny explains how other people can be a pleasant catalyser of compulsions, but "if other people don't participate" in compulsive interactions, "the effect is not as strong (...) and then I stop doing them, because that person doesn't share in the energy".

A more harmful co-presence of another person was articulated by Mina. In part, because she had to share her office with a colleague who had different preferences as to the appearance and use, she could not work to the desired standard anymore. The level of low light, particular door opening, closed cabinets, low phone volume, empty walls and desk and special chair provided her with what she recalled as "a truly luxurious situation". Her new office partner, on the other hand, required much more light, the door open, open cabinets, held meetings in the office, had put up many photos on the wall and small paraphernalia on her side desk. The presence of this colleague and her preferences completely rearranged, and thus destabilised the office up to the point that Mina had to change office. Her colleague's ownership of the different elements in the office did not change this, in particular because there was no distinct spatial boundary between their desks and the lighting and meeting activities enveloped the whole office space. In this manner, through the presence of other people, stabilisations are difficult or even impossible to maintain.

Other people can also become part of stabilising effort in a generative way. Lowri argues that she does not 'suppress' her compulsive interactions around people who she was "very familiar" with, such as her mother and partner. When she was younger, her "mum participated in the bedtime rituals" in which her mum's body, for instance, had to be touched in different places and her body had to be positioned in just-right places when Lowri brushed her teeth. Not only are other people part of compulsions, but they can also become part of stabilisations, and thus part of certain spaces for their willingness, mobility, and 'malleability'. Nonetheless, the presence of other people with whom Lowri was less familiar could take away the requirement for compulsive interactions and thus stabilisations. Her 'bed-time rituals' took so long that "the neighbour was brought in [to put Lowri] to bed because with her, it *did* work". With the neighbour's body physically there, she "quickly went upstairs", and because of whom such "evening was one I was at ease at". The neighbour might have neutralised the ongoing situation by potentially taking away all grounds for compulsive configuration of all that produced the spaces they moved through. In effect, any sense of necessity for stabilisation will have been disrupted as the social presence of the neighbour skewed all object excess towards the social meaning of objects.

Whilst the spaces up until this point in the chapter have mainly involved private spaces that became reproduced through their adaptability, other often less private spaces can be considered to be more rigid in their adaptability and duration. Nonetheless, any compulsive situation that can be stabilised will require compulsive interaction. For instance, when Bill turns his car into the small parking lot next to his house that consists of two rows of spaces, three out of the four spots in one row are taken, which requires him to "complete" the row by parking in the fourth spot. The requirement for compulsive interaction and stabilisation of a situation in such a rigid place can be difficult to negotiate because such an interaction is impossible or can be regarded inappropriate to varying degrees and kinds. Prevention of compulsive interaction in itself therefore seems to require negotiations by the human involved, as Sage ponders:

> Everything enters but I don't register it, and that's why it took such a long time for me to realise that I'm suffering from it (...) I think 'why am I so tired every time I get out off the bus?'

Becoming confronted with spaces that are unstable and compulsively require stabilisation then also provides a possible explanation for the experience of easily being overwhelmed by the absolving sights, sounds, and tactile that form the often rapidly changing world.

Through their rigidness and perhaps the lack of ownership or responsibility over such spaces, the appearances of the objects with which such spaces are produced need to be negotiated in other ways. In these cases, compulsive interactions are performed as avoiding extracorporeal systems to destabilise, and often involve the adjustment of other-than-compulsive movements. The observations and eye-tracking sessions outside domestic spaces provided insight into how rigid, and therein oppressive, structures become stabilised, neutralised or in some cases denied. The white stripes of pedestrian crossings and the tiles guiding blind people come to appear as structural stability when Sara crossed the road to the railway station. Only the white paint stripes are stepped on, despite it altering Sara's natural gait, and her path along the platform is determined by the ridges of the guiding tiles for visually impaired people. Unwilling or unable to challenge this system, she treats it as a given and performs in accordance with it. If she would not, her body might destabilise it, even if this would only be for the duration of a step.

Rigid unstable structures that cannot be stabilised through compulsive interaction often seem to leave gaps that then do require stabilisation for compulsive interaction. The conveyor belt of the counter in supermarkets presents a rigid rectangle within which groceries of different size become placed, and Ginny compulsively places one item after another with the widest side to the sides of the belt. In a compulsive stabilisation effort with Dylan, same-sized cartons of milk and yoghurt are placed on the belt parallel to each other. When putting the items in the boot of his car, he keeps grocery groups based on appearance in the same spot. For instance, eight packs of chewing gum that were reordered with the same sides up. As such, stable mobile systems consisting of non-human objects with shared dimensions then helped to negotiate larger rigid systems.

Some situations require compulsive interaction so strongly, and reconfigure the human body with such vigour, that it takes a lot of effort and active reconfiguration towards other-than-compulsive engagements to refrain from interacting compulsively. For instance, when Ginny sits in the passenger's seat, she can reach almost everything in the front of the car, but she crosses her arms and presses them tightly around her body to make it physically impossible to reach out and interact to resist configuration[5]. She pushes her thumbnail in the flesh of her other fingers to keep her hands occupied. Her response in effect demonstrates how the car immediately outside her restricted bodily reach becomes a space that poses challenges in its density of overlapping compulsive systems, which constitutes an incredibly strong pull towards her compulsively interacting with different elements of the car, my jacket as I sit next to her, or her own clothes, jewellery and her bag at her feet. Residing in the inescapable plethora of parameters of varying stability can only be negotiated by actively retaining a complete other-than-compulsive configuration of the body that requires a lot of energy.

Combining insights from compulsions emerging from stabilisations of body-world formations through systems of object-excess that becomes managed within parameters that people have differing authority over, provides new insights into the production of immemorial and non-human, but intimately felt time. Indeed, compulsivity demonstrates how these different formations have different internal endurance and that they can be managed in a variety of ways to extend their endurance. The different durations exist alongside each other; it is not one that is contained within objects, materiality as such, nor within the human body but is suspended in-between. Nonetheless, their inherent ties to the body's sensory registers and objects' physical form formation and capacities to disrupt emerge from different just-right realms of durations clashing with each other and disrupting other-than-compulsive durations. Compulsions then become reconciliations between these clashes and bring durations together.

In conclusion, corporeal environments reproduce an evolving ecology of a collection of non-humans with or without the body that stabilise or destabilise, which is supportive of Bennett's (2004, 2010) lively compositional ecologies. Compulsive interactions can be argued to 'mend' and/or retain the stability of these collections in order to enable performing other-than-compulsive life. The parameters within which objects, space elements, and bodies have relative 'freedom' to remain configured as just-right then further informs the ways in which turbulence (Serres 1995, Anderson and Wylie 2009) and friction between objects emerge (Deleuze 1991 [1988], Bennett 2001). Rather than objects and space elements causing turbulence and friction, this study suggests that the crossing of boundaries of compulsive parameters can also constitute these tensions. Such an understanding of bodily spatiality then further informs the geographies of medicalised performances similar to compulsivity, such as stimming, mania, and compulsions that are associated with existential fear (e.g. Davidson 2003, Segrott and Doel 2004, Davidson and Henderson 2010, Chouinard 2012).

Inescapable connections to ecologies of compulsive compositions help to explain Bliss' (1980), Kane's (1994), and Sage's experience of perpetual urge sensations, and thus the workings of the landscape of tension. Also, *not* compulsively interacting 'in line with' with extracorporeal rigid stabilities may invoke a compulsive 'crisis'. Derived from work from Dewsbury (2000), Harrison (2009, 2015), and Dawney (2013) that queries what remains outside or between events, this situates compulsions as opposites of crisis events or crisis management, precisely to counter the spontaneity, rupture, and interruption they produce. Remarkably, herein, compulsive interactions have made a 180-degree turn from being regarded as problematic acts and symptoms of a pathology that captures suffering to interactions that help to manage a socio-material situation to refrain a person from becoming overwhelmed. With the ecology of stable systems that shape and operate in parameters, compulsivity thus seems to actively produce space that orders bodies and their surroundings in accordance with their dimensions.

As compulsions should be considered to add to a minimisation of disruption and produce an ephemeral but lasting milieu that allows one to create a sense of well-being in a place, compulsivity's uncomfortable, disrupted relation between compulsions and the self should be revised. This relation can be unpacked and re-envisioned by employing Alphonso Lingis' (2016) argument of the location of the self. This starts with re-envisioning rationality, purposiveness, and meaning as signifying the self as a well-established and historical given. Lingis argues against the existence of a sense of self as an individual source of thoughts and decisions. Instead, these are always made against externalities, and the self emerges with them. As such, rationality, purpose, and meaning are just as situationally constituted as compulsions. Indeed, whereas compulsivity seems to focus on what is immediately in front of one, these concepts name an ordering of other domains of life (e.g. a social issue, grocery shopping, applying for a place for one's child at primary school) as a weighing of options that play out in abstract faculties. The former is traditionally not considered to be part of the self, as it holds no connection to these abstract faculties, the latter is. This is not to say that compulsivity should be considered as part *of* the self. Following John Wylie (2010: 102) compulsivity does not interrupt a "persistence of an undisturbed humanism. I mean by this the persistence of beliefs in the inviolate, coherent and given existence of a free-standing 'creative' subject – an undisturbed 'I' who feels, speaks, expresses and so on". Rather, compulsivity may instead be considered to be *for* the self. The next chapter explores how compulsivity is woven into everyday life and how the felt knowledges of the ecologies become mediations.

Notes

1 Compulsively seeing in this case and in many others is looking at something with intent and in full focus.
2 For instance, corners, curves, ceiling and flooring details, marks, spots, scratches, and other visual pattern disruptions.
3 For further reference on the percept, see Massumi (2002), and in human geography the percept has been employed in work on Non-Representational Theory approaches by Dewsbury et al. (2002), Wylie (2005, 2006), and Woodward (2016)
4 It is a room that resembles a 'box bed'; a bed that is immediately surrounded by walls.
5 This is also the technique Habit Reversal Therapy seeks to normalise to adopt when people want to refrain from ticking.

References

Anderson, B. and Wylie, J. 2009. On geography and materiality. *Environment and Planning A* 41, pp. 318–335.
Ash, J. 2015. Technology and affect: Towards a theory of inorganically organised objects. *Emotion, Space and Society* 14, pp. 84–90.
Beljaars, D. 2020. Towards compulsive geographies. *Transactions of the Institute of British Geographers* 45 pp. 284–298.
Bennett, J., 2001. *The Enchantment of Modern Life: Attachments, Crossings, and Ethics*. Princeton, NJ: Princeton University Press.
Bennett, J. 2004. The agency of assemblages and the North American blackout. *Public Culture* 17, pp. 445–465.
Bennett, J. 2010. *Vibrant Matter: A Political Ecology of Things*. Durham, NC: Duke University Press.
Bergson, H. 1911 [1896], *Matter and Memory*, trans Paul, N.M. and Palmer, W.S., London: George Allen and Unwin.
Bliss, J. 1980. Sensory experiences of Gilles de la Tourette syndrome. *Archives of General Psychiatry* 37, pp. 1343–1347.
Chouinard, V. 2012. Mapping bipolar worlds: Lived geographies of 'madness' in autobiographical accounts. *Health & Place* 18, pp. 144–151.
Colebrook, C. 2002. *Deleuze*. London: Routledge.

Cresswell, T. 1996. *In Place/Out of Place. Geography, Ideology, and Transgression.* Minneapolis, MN: University of Minnesota Press.

Davidson, J. 2003. *Phobic Geographies.* Aldershot: Ashgate.

Davidson, J. and Henderson, V.L. 2010. Travel in parallel with us for a while': sensory geographies of autism. *The Canadian Geographer/Le Geographe Canadien* 54(4), pp. 462–475.

Dawney, L. 2013. The interruption: investigating subjectivation and affect. *Environment and Planning D: Society and Space* 31, pp. 628–644.

Deleuze, G. 1991 [1988]. *Bergsonism.* New York, NY: Zone Books.

Deleuze, G. and Guattari, F. 1994. *What is Philosophy?*, trans. H. Tomlinson and G. Burchell, New York, NY: Columbia University Press.

Dewsbury, J.D. 2000. Performativity and the event: Enacting a philosophy of difference. *Environment and Planning D: Society and Space* 18, pp. 473–496.

Dewsbury, J.-D., Harrison, P., Rose, M. and Wylie, J. 2002. Enacting geographies *Geoforum* 33, pp. 437–440.

Harman, G. 2002. *Tool-being: Heidegger and the Metaphysics of Objects.* Chicago, IL: Open Court.

Harman, G. 2018. *Object-Oriented Ontology A New Theory of Everything.* London: Pelican Books.

Harrison, P. 2009. In the absence of practice. *Environment and Planning D: Society and Space* 27, pp. 987–1009

Harrison, P. 2015. After affirmation, or, being a loser: On vitalism, sacrifice, and cinders. *GeoHumanities* 1(2), pp. 285–306.

Ihde, D. 2009. *Postphenomenology and Technoscience.* Albany, NY: SUNY Press.

Kane, M.J. 1994. Premonitory urges as 'attentional tics' in Tourette's syndrome. *Journal of the American Academy of Child and Adolescent Psychiatry 33*, pp. 805–808.

Lingis, A. 2016. Aconcagua. In: Wheeler, R. ed. *Passion in Philosophy: Essays in Honor of Alphonso Lingis.* Lanham: Lexington Books, pp. 3–16.

Manning, E. and Massumi, B. 2014. *Thought in the Act: Passages in the Ecology of Experience.* Minneapolis, MN: University of Minnesota Press.

Massumi, B. 2002. *Parables for the Virtual: Movement, Affect, Sensation.* Durham and London: Duke University Press.

Segrott, J. and Doel, M.A. 2004. Disturbing geography: obsessive-compulsive disorder as spatial practice. *Social and Cultural Geography* 5(4), pp. 597–614.

Serres, M. 1995. *Genesis*, trans. James, G. and Nielson, J., Ann Arbor, MI: Michigan University Press.

Simondon, G. 2020 [1964]. *Individuation in Light of Notions of Form and Information,* trans. Taylor A., Minneapolis, MN: University of Minnesota Press.

Woodward, K. 2016. The speculative geography of Orson Welles. *Cultural Geographies* 23(2), pp. 337–356.

Wylie, J. 2005. A single Day's walking: narrating self and landscape on the southwest coast path. *Transactions of the Institute of British Geographers* 30, pp. 234–247.

Wylie, J. 2006. Depths and folds: on landscape and the gazing subject. *Environment and Planning D: Society and Space* 24, pp. 519–535.

Wylie, J. 2010. Non-representational subjects? In: Anderson, B. and Harrison, P. eds. 2010. *Taking-place: non-representational theories and geography.* Farnham: Ashgate, pp. 99–114.

8 Mediations

Finding ways with compulsive life

Compulsive lives are spatial lives on many different levels. The previous chapters develop the anticipated constitution and enduring persistence of the various aspects of compulsions as felt and known through the spaces one dwells in and the objects one perceives and handles. There is an experiential knowledge that emerges through learning that the relations between the body and its surroundings can be mediated. Stabilisation of compulsive systems and organisation of compulsive compositions seem to be an important part of the efforts to diminish the interruption from interactions and getting stuck, as well as increase moments of enjoyment. As these efforts are already forms of mediation, this chapter explores further how people employ everyday life strategies to restrict configuring towards compulsive situations, how bodies, objects, and spaces could be organised to further stabilise certain spaces. To this end, it critically reflects on the study of compulsion in Tourette syndrome and the principles that have guided support and treatment currently on offer for this group.

Habits and durations

Can there be any respite in the spatial turmoil from which compulsions emerge? Given they are imbued with so much compulsive potential, can relaxation be permitted in the spaces of the home and other important places that are so important in other domains of our lives? Dylan, who spends a lot of time at home[1] describes how he experiences being calm in the context of compulsivity and Tourette syndrome more broadly as "I am on earth, I am here". Echoing similar descriptions from other participants, it is distinctly immediate and situated; it conveys a groundedness and a feeling of being completely at ease with his surroundings. His peacefulness exists amongst and in-between the compulsive compositions and parameters. Perhaps remarkably, he does not immediately dismiss compulsivity as precluding feeling at ease, despite being one of the few participants who would rather not do any compulsions (including tics). Indeed, Ginny argues that compulsivity is intricately part of finding everyday peace:

DOI: 10.4324/9781003109921-9

The art of the matter is that you have control over what compulsivity you allow or not, and that works for me, quite well actually.

For her, it is more about finding the right balance of allowing configurations of the body in compulsive enlacing to unfold with resisting other compulsive configurations. This balancing act to feel at ease seems to continually evolve, which is in part derived from the instantaneous social context, rather than a planned strategy that is reasoned through. She had been asked by her therapist if she could 'hold them up' to 'release' them in another situation:

Yeah, well, I don't know and I did want to try is because I have ... – the other day in the car when no one else was around I thought to give it a go. But then things come out that surprise me, like 'I never do those!' (...) When I was younger, I'd go to noisy concerts of those rock bands, because I could just let go, because that was actually really pretty good!

Considering living peaceful compulsive lives as emerging with the ecologies of more or less stable systems of multiple body, object, and space elements, two realistic possibilities are conceivable if we discount interference with the body's sensory capacities (i.e. medication), tolerance to persevere through sensory discomfort (Exposure and Response Prevention), and curtailing of movement (Habit Reversal Training). These are based on the spatial systems.

The first way to diminish the disruptions compulsions can cause to other-than-compulsive life but retain the enjoyment of just-right environments entails residing in environments that consist of either very few or very stable systems that also do not destabilise each other. Such environments require very little to no compulsive intervention for an extended period. Ginny presents an image of "large grasslands that are not fenced", with Kyrgyzstan as "ultimate spot" because there is "nothing at all, just green hills, you know, where there's nothing in sight". She also likes the sea, and the desert for being "nice and calm", although it is "too noisy because of the wind there". Moreover, for the high expectation of remaining compulsively unperturbed, Alan likes taking a stroll in cemeteries and is fond of walking in the rain because "that reduces stress better". Siôn reminisces over living in Zeeland[2] as he grew up on a farm and is keen on moving back because he finds it "a nice place, but mainly [because of] the peacefulness; the expanse of the polder[3]". Also,

The church bells play in a particular way. I became very tranquil when I heard that sound. Then I think, 'oh man, I'm in another world'.

As mentioned earlier, he refers to environments that he has organised to be just-right as particular worlds or planets. In the house he lived during the study, he liked residing in warm, light, sunny places, and especially the home-office room he designed and constructed incites him to …

> … just stand there, like, very quietly, then I could look for hours. Bizarre really, that you can just enjoy a piece of tranquillity (…) I can kind of explain it as that you feel so happy that your eyes are at peace again.

In the brightness of the room, the denotation of a tranquillity emerging from his eyes being "at peace" suggests how the room holds a visual absence of incitements of compulsivity. In contrast, Ginny needs to reside in a darkened room when she feels overwhelmed, for instance after a busy day at work: "after coming home, I first need to be in a dark room, curtains closed" to "meditate". "After an hour or some I could do cooking or something". Whereas being in a bright white room to reduce compulsive tension could be considered to be oppositional visuals, in compulsive terms, they are strikingly similar in the sense that both spaces become a blur of colour in which very few objects and space elements stand out. Indeed, this also reflects Mina's living room in which she cannot tolerate bright colours. Spaces that lack much distinction in terms of colour consequently also lack distinguishable compulsive compositions or have very few differing parameters of potentially compulsive collections. As such, from a compulsive perspective, dulling colour difference in a space incites a therapeutic feel to such spaces. Whilst such assertion is not necessarily new in the literature of therapeutic spaces and landscapes (e.g. Gesler 1992, Wilson 2003, Conradson 2005, Lea 2009, Bell et al. 2017, Gorman 2017), as especially people experiencing episodes of fear, anxiety, mania, and other types of stress as well as numbness and lethargy, tend to find that residing in dark spaces help them to recuperate (Lipsky 2011), insights from compulsivity may contribute to explaining what constitutes this effect.

Whilst remaining in spaces that reduce the need to perform compulsions is unpractical for most requirements of everyday life, or accept strict colour and design schemes, there are other ways of residing in a plethora of spaces and avoid becoming enlaced in compulsive interactions. These are captured in the second way to live peacefully with compulsivity based on spatial systems. It entails the elongation of system stability in places. In addition to holding objects in place and assigning them their own spot, as explored earlier in this chapter, such efforts also find support in other-than-compulsive interactions. As maintaining stabilisations of body-object-space collections helps to postpone compulsive interactions that re-stabilise collections that have fallen apart or are threatening to do so, postponing other-than-compulsive acts that could destabilise these collections would mean causing as little disturbance to stabilised situations as possible. In addition to staying seated in one place, as Bill and Dylan do, doing everything in the same manner; handling objects in the same fashion, making the same movements, approaching furniture from the same angle, referring to the body in similar ways, maintaining order and timing then reduces the chance to encounter situational change. Indeed, many people with compulsive sensibilities tend to routinize, habituate, and carefully plan the very practical dimensions of their lives. Ginny elaborates:

> If I thus have a very structured plan, then that is a 'thick current' with which I go along, like 'I'm going to do this and I'm going to do that', and then I suffer less.

Such routinized ways of going about everyday life conjure up a highly embodied future that consists only of the gestures and moves of other-than-compulsive futures. These kinds of imagined futures help in retaining the bodily configuration to other-than-compulsive aspects of life. When Elisa works in the garden, she follows a particular order, which makes the task ...

> Elisa: ... very predictable and very calming, because I know that then and then, I'll do that and that, and I take a break to do the floor because it just irritates me, then it's ok, if I make sure to continue with the same order afterward. (...) With cleaning, I experience much more compulsive tension; things must always be more precise and straighter, and that's ... exhausting. You kind of want to let go of it, but you don't want to let go of it, because then things aren't right, and that's very annoying. And that's not the case here. It can never be a 100% correct, so it's just nice being busy with nature and that little order and just nice and relaxed ...
>
> DB: yeah, that it becomes possible to ...
>
> Elisa: To be away from myself a while.

As Elisa's argument suggests, habits, routines, and compulsive interactions appear so strongly entangled that they can become indistinguishable. This is shared by Bill as he explains that brushing his teeth in the morning is so strict that it is compulsive. The routine has become so particularly organised that when anything is out of order, for instance when the toothpaste is not where it should be, or when the electric toothbrush is not charged properly by family members, he cannot finish the routine. The series of other-than-compulsive acts and the perceptions that come with them have in their serial performance rigidified in percepts that indicate a just-rightness. When anything is out of order, Bill's distress does then not only emerge from having to locate the toothpaste or wait until the brush is properly charged, it is also underpinned and exacerbated by the impossibility to make it just-right, unless the whole ritual is started afresh. However, when all is in place he can brush his teeth going seamlessly from one just-right motion and perception to the next. The routinisation of everyday life tasks is a well-known strategy to diminish various kinds of anxieties to emerge from unexpected occurrences, particularly well-developed by autistic people and those with ADHD and OCD sensibilities in addition to those suffering from phobia and mania (e.g. Davidson 2003, Lipsky 2011, Chouinard 2012, De Caluwé et al. 2020).

This serialisation of compulsions into other-than-compulsive activities has seeped into more aspects of Bill's life. Before taking a sip from a glass, he pinches the sides of the glass twice, and squeezed it before putting it down. This and other elaborate compulsive interactions that have a seemingly habitual element to them require him to take his time, but they are unshakable; he cannot do them when is convenient to him. Bill and others also argue that certain compulsive interactions *have* to be performed *before* other-than-compulsive interactions. Almost every time Tomos engages in other-than-compulsive task that requires him to use his hands, he rubs his face beforehand; for instance, before starting his laptop and before picking up his cutlery to eat. Whilst these are relatively short and 'easy' compulsive interactions, Mina requires a complex and elaborate series that needs to be exactly right before she can close her eyes and sleep. She calls this series a "mandatory ritual", and if anything "goes wrong", she needs to "start all over again", which includes going to the bathroom. Compulsion thus precedes habit, as compulsions invoke a habit because it incites a strong preference or rigid manner of action, which can be a means of breaking the compulsive enlacing and reconfiguring towards other-than-compulsive life. For instance, before Sara vacuums the upstairs room, she looks at the bedside clock until the minute passes, because she times her vacuuming to exactly seven minutes. If she would not count the minutes, she would have to continue vacuuming and face difficulties in reconfiguring to stop. Habits can thus become mediations of compulsive interactions. By extension, the merging and co-constitutive emergence of compulsivity and habituality support the elongated durations of stabilised situations through other-than-compulsive interactions and activities more broadly.

This productive intertwining of compulsivity and habituality has consequences for how habit formation and organisation is understood. It moves away from approaches to habit through functionalist practice theories (see Schatzki 2008), in which they are posited as different, almost mechanistic, method of completing tasks, rather than meaningful reasoned action. Instead, it inches closer to vitalist understandings of habit. When Elizabeth Grosz (2013: 217), drawing on Ravaisson, Deleuze, and Bergson, argues that "[i]n a world of constant change, habits are not so much forms of fixity and repetition as they are modes of encounter materiality and life", it resonates with compulsivity. Indeed, as many compulsive interactions recur in simultaneity with habitual engagements with the world, the formation of habits could be considered to be underpinned by compulsivity[4]. In addition to the action beyond reflexive thought that habits constitute (Bissell 2011, Dewsbury 2011, Dewsbury and Bissell 2014), the vitality, creativity, and futurity are revealed in the compulsive mode.

In short, negotiating compulsive tensions thus also takes place in a distinctly spatial manner, using the more or less rigid organisation of bodies, objects, and space elements to reduce disruptions. Additionally, in compulsive terms, places acquire a certain quality on their potential to induce compulsions based on the prevalence and stability of compulsive compositions, as well as the familiarity, sociality, and the kind of other-than-compulsive activities that tend to happen there. Furthermore, mapping anticipation of unstable compositions is possible through the adoption of habitual movements. Compulsivity as spatial phenomenon may also contribute to explaining why 'cluttered' environments require more attention when engaging in an intended practice (see Land et al. 1999, Hayhoe et al. 2003, Belk et al. 2007, Arsel and Bean 2013, Dion et al. 2014, Owen 2018). The potential for becoming reconfigured into compulsions is then very high, and navigating the systems without having to stabilise, restore, or disrupt any is, in Elisa's words, "exhausting". By returning to Lingis (1995: 602), who argues that "we see not shapes but possibilities" (see also Bergson 1911), it may be possible to see how this formulates the struggle by people with compulsive tendencies. The possibilities for body, object, and space dimensions to form an unstable compulsive composition are then oppressive in the compulsive interactions they evoke. The anticipation of compulsive interactions then infers navigating the ecologies of these possibilities by reducing them.

Compulsive interactions become a response to an ever-changing ecology or patchwork of different kinds of intensities and with differing durations, that articulate the particularity of the situation through the similarities of the bodily, object, and space that constitute compulsive configurations. This renders compulsivity as a disposition of the corporeal enlacing and simultaneous violence to retain a sense of corporeal unity. Compulsive interactions can then be understood as confirmation of the fleshiness and presence of the body in relation to the materiality of the extracorporeal element. Simultaneously, they can be understood as slippage from human governance and rupture from the body's assumed raison d'etre; its sustenance of life – biological, cultural, social. Compulsive interactions name the ecologies of intensities that enmesh situations of the human and non-human attuning to a plane of consistency with affects that meet and explode in a multiplicity, reverberating through the human body.

On the compulsive organisation of other-than-compulsive life

As the spatiality of compulsivity provides insights in the constitution of individual compulsions, this spatial understanding also suggests how to create and reproduce worlds that do not evoke compulsions as vigorously as they can be in certain environments. However, falling into the trap of recreating a spatial determinism that underpins many assertions that push located and territorialised well-being is a real possibility here. Rather than restating age-old arguments that argue that 'green spaces' and natural areas are good for people, compulsivity demonstrates how such spaces tend to be associated with diminished not-just-right experiences, which is a dimension that goes beyond meaning, sociality, and their potential for some physical activities for some people. Simultaneously, it demonstrates how such spaces are not unquestionably *good*, as not-just-rightness is still present in these spaces, as if shifts patterns of lines, colours, temperatures, textures, etc., that emerge from indoor spaces and objects to patterns at the scale of landscapes, vegetation, and man-made structures and objects. Staying faithful to the situational ontology on compulsivity and its configurational pull on bodies, objects, and spaces, what follows is an application of the compositional principles to avoid compulsive interventions altogether and those that increase the duration of stabilisations. These mediations are grounded in the experiences of the research participants and have been theorised to fit situations that are similar on compulsive terms.

Offsetting perturbance that comes with moving with the landscape of unrelenting tension that is never really dies down is based on slowing down the intensification of tensions. As mentioned, this translates most strongly into dwelling in places that tend to be more or less universally associated with less compulsive interventions, such as outdoor places with vegetation, natural areas, parks, and gardens in particular. The presence of plants and other 'natural' elements for decoration, such as plants or pieces of (unprocessed) wood, may also help to reduce tension in that corner of a space, or add just-rightness that can be enjoyable to be around.

Residing on one's own in one spot for a long while in a space could help as the position of the body remains the same and the chances to intersect with percepts that indicate a (not-)just-rightness are lower than when active in a space. Changing light conditions do change the organisation just-rightness of the space as time passes, but this may be minimal. This should be caveated that residing in one spot but looking at the details of the place will have opposite effects as with increasing familiarity with the details of object collections or wallpaper pattern more (not-)just-right patterns can arise. As such, focussing one's senses on one thing takes away possibilities of new patterns arising as the background fades in one's perception. The thing focussed on may see new (not-)just-right patterns emerging *unexpectedly* which diminishes the disruptive element. Visually, by looking at a screen patterns tend to be minimal, and in terms of touch through crafting or making (as Mina does), the emerging patterns may be intentional and therefore enjoyable.

Stabilisations can also be organised through making anticipated situations as predictable as possible. As reflected in the practices of many research participants, routines helped towards avoiding disruptive bodily configurations. Highly detailed knowledge about situated objects and task performances (e.g. making and having breakfast) also helps to recognise when certain objects and spaces can become 'demanding'. For instance, being able to reach for the butter in the fridge with one uninterrupted movement and without having to search for it would keep the threat to be configured at bay. Nonetheless, precisely with detailed knowledge of the various aspects of the routine building up, new compulsions can take shape. Hence, there is a fine balance, but a distinct difference, between what Mina calls an 'innocent ritual that is more like a habit' and a 'ritual with a mandatory character'. For her, the latter category started as a lighthearted routine, but grew into a serial compulsion she has to complete just-right, or start over the routine in its entirety.

Mediating the landscape of tension can also work utilising sociality. Following Lowri's experience in her teenage years, the accompaniment of another person, in particular one who is relatively unfamiliar, during routine tasks that tend to conjure up many compulsive situations seems to work in a desensitising manner as their presence requires a lot of attention and pre-occupation. The latter puts more pressure on the unfolding of the task, which takes away the ease with which the body would otherwise be configured compulsively to a range of interactions that involve tactility, auditory just-rightness, and visual ordering. This is not to say that such a person should be around at all times, in fact, that would increase incite a shift from primarily task-consciousness to primarily body-consciousness which is likely to result in putting energy towards intentionally reducing the visibility of compulsions and tics rather than completing a task. In addition, echoing Elisa's and Alan's knowledge, associations between other people and objects and spaces may help to mediate ordering compulsions. If ownership is particularly clearly of another person, compulsive attendance seems to be required to a lesser extent. It could be speculated that this hesitation has to do with the current order that could be just-right for that other person and from an angle the individual comes across the setting. Furthering just-right logic, the imposition of 'one's own' just-rightness could disrupt the current just-rightness, and that is unacceptable.

Diminishing the peaks in the landscape of tension and keeping the difference between the peaks and lower levels as minimal as possible to avoid differences in a space, would be possible by initiating monotone bodily sensations keep a certain rhythm going. Such a rhythm that is produced by one's own body and helps to 'keep it together' 'overshadows' rhythms that emerge from changing bodily environments as these may cause flare-ups in the landscape of tension. As such a bodily rhythm dampens the effects of differential rhythms of the extracorporeal, the body becomes less vulnerable to be configured towards a compulsive situation. Monotone sensation can be produced by monotone activities, for instance, Sara always has her two small stuffed animals to hand, and in particular in situations that she is more vulnerable to be swayed by her surroundings to perform a compulsion, she rolls their seams through her fingers (see Chapter 4). Further reducing the possibilities to become enlaced into a compulsive situation by keeping her hands occupied, it closely resembles stimming, the calming activity of autistic people do. The differential rhythms on which compulsivity is based may therefore also provide new insights into the success of stimming. Furthering this argument may also suggest why music therapy is considered a successful form of therapy for Tourette's (Bodeck et al. 2015). Music might help in bringing harmony in a space, as it is an auditory incitement to move with that works through the body, and as a rhythm that may drown out other rhythms and introduce a more acceptable version of just-rightness in a space.

To retain a restorative element in one's house, room(s), or other spaces that allow for being adjusted, there are different ways to alter the appearance of spaces and constitute the organisation of the body in a space that are likely to diminish the compulsive 'pull'. When things are just-right from multiple angles, in different lighting conditions, and in tactile, visual, and auditory sense, they could introduce a just-rightness to a room or a part of a room if they are placed in a visual context that exacerbates the just-rightness, or in the very least, does not disturb it. For instance, Mina meticulously woodworked a decorative chest that she had carved into a just-right state, and which she has on display as visible from almost her entire living room. Also the reverse may be true; objects with dynamic details like fringes on a carpet or chair can be anticipated to move or have moved when doing other-than-compulsive activities, which presents a tension and incites compulsive checking and stabilisations.

As compulsivity tends to work towards stabilising body-object-space formations, compositions that last longer are preferable. Locking items in place as Elisa does, assigning mobile objects to specific spots as Joe does, and putting items out of sight as Elisa does are all viable strategies. In the same fashion, other stabilisations could entail grouping items, like putting things on a tray, in a basket or box, or in a bag helps 'contain' these things together and prevent the system from falling apart. Also, as Ginny does, entities can be kept from *moving* by putting the same kinds of products/objects in close vicinity. Such entities would not need to be similar in function, such as all jam jars, but similar in object dimensions, such as having similar sized spherical shapes, have similarly long vertical lines and/or materials. Sports bottles could, for instance, be grouped with shampoo bottles.

Avoiding having to *move with* shapes, temperatures, and textures becomes possible in several ways. For instance, having to poke or press in or run one's nail along small ridges and bumps, such as certain types of (old-fashioned) wallpaper and rough or untreated wood is particularly challenging. If these textures and objects cannot be covered up or removed from familiar spaces, textures could be made less apparent by placing it behind a glass plate, paintings for instance. A glass cover effectively erases, or at least diminishes the appearance of the texture, and it makes the texture that is still visible impossible to reach, which, in turn, diminishes the requirement for interactions. Moreover, as not-just-rightness also tends to emerge from 'dramatic' primarily visual and auditory changes, such as lines between opposing colours and expressing sharp edges of objects against a different background. Large checkered floorprints that can be found in UK and Dutch kitchens are prime examples to avoid. Instead, flooring such as carpet, tiles, and linoleum that have gradual colour changes would invoke less compulsive tension than those with defined lines and (large) colour blocks. Better still may be flooring with a natural grain, such as cork or a parket floor or wood print with minimal colour range and no dramatic colour difference.

As compulsions that invoke pressing a body part into an object emulate the dramatic nature of form, for instance sharp edges, tips, heat, but also velvety textures, hard slick surface, and soft bounciness of the flesh may be mediated through an alteration of these forms. Therefore, with the introduction of new items in a space, such as the home, workplace, or place of study exacerbated forms should be avoided where possible. Also, in terms of alteration of objects in one's ownership, sharp corners of large surfaces, such as walls or tables, could be rounded off to a *certain* degree. To the certain degree in the sense that it takes away the dramatic form, but also that the new edge does not immediately correspond to the various forms of the hand, such as when cupping, the space between two fingers, or the size of a phalanx. If alteration of form is not possible, placing items against similarly coloured and textured backgrounds dims the dramatic appearance of the form or colour that stands out. Furthermore, avoiding not-just-right sounds when possible could work with wearing earplugs or playing music through headphones. This especially reduces the anticipation to be confronted with not-just-right sounds to acceptable levels in public places. In domestic spaces, identifying what makes particular sounds that is not-just-right, and attending to it may be worth it. In addition to wearing headphones (with or without music on) indoors, for instance grease door handles and hinges, shaving off a couple of millimetres from the underside of a door, and attending to the sound of kitchen appliances before buying it might be helpful. This exploration of what might work to mediate being configured towards compulsive situations from a situated perspective starkly differs from existing, formal structures of support and treatment. Of course, these suggestions are not on the same footing as these structures, however, the underlying principles do inform the current support and treatment options, as well as the practice of Tourette syndrome research.

On the study and treatment of Tourette syndrome

The most important shift with reference to Tourette syndrome in the debate that the argument so far has initiated is that the interactive dimension of the array of compulsions ought to be considered as originating in the relations between the body and its surroundings rather than *only* in the brain and further nervous system. The corporeal environment should not only be seen as *influencing* the manifestation of compulsions, rather, it should be treated as *active constitutive component* of compulsivity. Indeed, the demonstration of the *fundamental* involvement of the extracorporeal in the production of compulsive interactions, it should be considered on an equal footing as the brain. Whilst this assertion is potentially less relevant for 'simple' tics that are not experienced as immediately interactive, it has several consequences in that it incites several fundamental expansions in major domains of Tourette syndrome, including the production and utilisation of knowledge.

Not by any means do these expansions call for an abandonment of certain types of research and support, rather, these expansions demand a reflection of what really matters in Tourette's, and more importantly, living in accordance with Tourette's sensibilities. They do infer that the research paths of the focus on the brain and cognition, that have been chosen decades ago, should lose some of their dominance as they currently have in the valuation of what constitutes the best knowledge of Tourette syndrome, and how profoundly this knowledge feeds into pathways to formal support systems. Although it sheds perhaps painful new light on the limitations of styles of inquiry that govern the materialist medical and clinical sciences (see Beljaars and Bervoets 2021), in particular the deficit models that frame their knowledge creation (Bervoets and Beljaars in review), there is clearly much to gain in broadening the scope, remit, and purpose of Tourette research and the ways in which findings filter into formal and also informal support systems.

In reflection of the gaps in scientific knowledges that were identified in Chapter 2, the reconstruction of compulsivity in Tourette's as also having spatial underpinnings opens up several new avenues of inquiry as well as new principles to take into account in the further study of the syndrome. In addition to those related to compulsions and complex tics, and various types of contexts, rethinking how to incorporate these insights start at knowledge construction itself. In particular, it cautions against valuations and acceptance of reductions of phenomena, categorisation efforts that are limited in their fit for the purpose of improving lives, inference of artificially produced claims to situations beyond the laboratory and research setting, and overclaiming both the role of the brain in the constitutions of compulsions and the usefulness of certain types of information for living with Tourette sensibilities. Indeed, the rendition of compulsivity as *behaviour* is precisely why there is so little knowledge about compulsions and tics in Tourette syndrome. As they cannot grasp the minutia, catch-all therapeutic interventions that approach compulsions and tics as Tourette symptoms utilising a behavioural lens will therefore always offer a partial diminishment of compulsions and tics[5].

The bodily, social, and material circumstances of individual compulsions uncovered in the previous chapters that produce a range of experiences, bodily responses, and social engagements call for the well-rehearsed hesitation to the neuropsychiatric, biomedical, and clinical scientific drive to produce correlative – rather than causal – answers to why-questions. Amongst other things, this pertains to 'waxing and waning', with which the biomedical and clinical sciences indicate the unexplained emerging and vanishes of a particular tic or compulsion over time. Their unsituated approaches only have a partial premise to understand this phenomenon, as they do not ontologically account for the context. This is also a reason why the rebound phenomenon is also fundamentally misunderstood. The situated theory developed here suggests that the build-up of tension in a space, that cannot be released through compulsions, may come out in a different context. That is precisely because one is presented with new evocations that had not hermetically been sealed off from engagement as what happened in the previous place.

Downplaying context is damaging; it limits the viable answers to those that can be entirely derived from the person, or more precisely the body in question. This reflects the knowledge production in laboratory conditions as it furthers the depiction of the realities of Tourette syndrome as playing out in a vacuum. Indeed, despite the poststructuralist and postmodernist critiques of knowledge production from the 1980s onward that point out the problems with linear models, pursuit of universalism, and endeavours that seek to stabilise phenomena in rigid systems, neuropsychiatric research in Tourette's has not taken this into its stride. As is notably pointed out in science and technology studies, actor-network-theory approaches, critical feminist theories, and postphenomenology (see e.g. Mol 2002, Law 2004, Latour 2005, Ihde 2009, Latimer and Miele 2013), methods, protocols, and technologies are not objective and detached from the reality they supposedly uncover (see also Wong and Beljaars forthcoming). Rather, they produce a version of the phenomenon.

Breaking with the kind of compulsivity and Tourette syndrome that had been shaped in these neuropsychiatric, biomedical, and clinical sciences, the social scientific approach in the current study does account for and welcomes multiplicities, nuance, and categorical heterogeneity. In the generative mode, 'horizontal' knowledge creation in which agency is distributed beyond the human, normativity is halted, and the ephemeral in addition to the material has a place in the analysis thus produces different versions of compulsivity and Tourette syndrome. Versions that are more recognisable for people with Tourette sensibilities, and versions that advocate for a shift towards *responses* to struggle and suffering. This study reminds how compulsions always *also* have a fundamentally social base, which points out that societies have a responsibility to address the suffering that emerges from Tourette syndrome. Being stared at when ticcing is a major source of suffering, yet no scientific efforts have been directed at addressing this. And whilst it is perhaps unfair to request this effort from the medical and clinical sciences of Tourette's themselves, endeavours to support these efforts should either have been an integral part of the studies or welcome social science and humanities approaches that include addressing suffering.

Moreover, these versions incite a hesitation to research that offers technological *solutions* to what is perceived to be the problem. Echoing Duff (2014),

Tyner (2016), McPhie (2019), and the broader critical approaches to the construction and organisation of conditions, health, and illness that have pointed out how in neoliberalism the processes of solution-driven problem definition in the context of mental illness and mental health disorders tends to preference outsider or 'healthy' interpretations. With reference to Tourette syndrome, solutions that become viable overly draw attention to the visible aspects and skew the responsibility of the necessary efforts to the person who performs compulsions.

Another fundamental shift advocated for on the basis of the book's argument is the theoretical grounds for support, which immediately raises questions on what constitutes good and appropriate support. Even treatment needs to be rethought. Indeed, there are now new reasons for attending to the bodily environment to diminish the disruption that compulsive interactions can instate. Such attendance may help to diversify the current range of formal support options that include chemical alteration of the biological body that changes the sensory registers, and incites profound further bodily changes that are conceived and accepted as side effects. The range also includes an offer of techniques of self-policing the noticeable bodily movements and ignoring sensations and feelings, which for many people goes against what feels as natural instincts that are not too dissimilar to common biological processes, and which, in turn, create ingenuine and harmful unemphatic or even conflicting relations with the body. Indeed, whilst there is a good range of addressing Tourette's as an individualised problem, the socio-spatial reimagination of the phenomenon requires an expansion to societal changes, starting at the social spaces of everyday life.

Caring more and differently and finding out how to care for people who suffer from compulsivity instead of trying to repair them is one of the central arguments Donna Haraway makes in *Staying with the Trouble* (Haraway 2016). It feeds into a welcome pivot into considering the organisation of the most helpful attendance to the place compulsivity takes up in people's lives. Whilst in clinical practice this is conflated, Haraway reminds us that the difference lies in the prioritisation of knowledge systems. Whereas Haraway in conjunction with disability advocates and the interpretative social sciences of disability, disease, and illness press for experiential knowledge to have a high priority in decisions on structure of support and treatment, clinical sciences prioritise cognitive and neurobiological knowledge to decide on their provision of treatment structures, which is informed, but not guided, by experiential knowledges. Such knowledge formation would allow more space to attend to the places in which people's lives play out.

The argument developed in this book conjures up a set of questions about how Tourette syndrome is understood and conceptualised, in particular to its status as 'disordered state of being'. The *meaning* of the Tourette syndrome diagnosis should name not a person's defect, but that way in which they can be affected or the kind of suffering that can overcome them because of specific situations, but which remains a ground for the provision of generous state support. A turn to the situated urge sensations and performance of individual compulsions interactions expands Tourette's as a particular affective connection with the bodily surroundings. These connections express what has remained non-representational and inexplicable, as the vocabulary necessary for these experiences has not been developed in formal medical and clinical contexts, nor do formal meetings for the discussion or delivery of treatment outside medical spaces. The latter prohibits a fuller understanding of how and which particular compulsions pose challenges and what elements of spaces and social realms are associated with difficulties. In addition, in failing to capture the precognitive affective dimension, explanations of compulsions through biological and historical psychoanalytical logic ultimately remain partial. Accounting for this dimension more fully may also provide a fuller and more generous understanding of coprolalia. Admittedly asserted in the speculative mode, appraising swearwords that are contextually sensitive through a lens of affective linguistics may indicate aspects of that situation that are not captured by signification and social language. Indeed, as a precognitive utterance it bursts out into the world *as if* it were cognitively considered.

The ways in which the bodily surroundings are co-constitutive of compulsions demonstrates a strong pull of people's attention to the here and now: the multitude of details of their socio-material situation are, as Sage phrases it, always and inescapably registered. To live more fully in the moment by retaining awareness of one's body and what is sensed, rather than being more fully swept up in more abstract social, political, economic, and cultural domains of life, may also underpin certain other struggles that have not been considered according to the same principles. For instance, the distress that presents as rage attacks seemingly emerge instantaneously, without reason, and with unexpected intensity, could very well have strong roots in the bodily surroundings and build on an overwhelming sense of not-just-rightness.

Notes

1 He did so also before the Covid-19 pandemic.
2 Southern province in the Netherlands, which is a coastal region that is open and flat and consists mostly of large-scale agricultural lands with small villages dotted in the landscape.
3 Polders are rectangular strips of drained land below sea level that give the landscape in the rural western and northern parts of the Netherlands a pillowy look.
4 I am leaving the suggestion open for exploration beyond the book, and a much more in-depth reading is needed to identify how and where thinking habit through compulsivity can inform both either conceptually or as lived phenomena.
5 Diminishing compulsions and tics does not equate a diminishment of suffering per se, and their conflation is problematic. At the time of writing this book, (some) resistance to behavioural therapy that is on offer for people

with Tourette syndrome is mounting. Behavioural therapies are likened to Applied Behavioural Analysis (ABA) therapy that had been developed for autistic people. ABA is severely criticised for its ideological goals and practical application that subjects people to 'techniques' that amount to abuse and torture in other situations (there is extensive critique in both scholarly and activistic circles, see Kirkham 2017, Rosenblatt 2018). In conjunction with broader 'neurodiversity' movements, Tourette's self-advocates (Daniel Jones and Luca Dolstra amongst others) argue that behavioural therapies share the same principles as ABA and promote Tourette's as neurological *difference* rather than *problem* and call for more societal acceptance and institutional accommodations.

References

Arsel, Z. and Bean, J. 2013. Taste regimes and market-mediated practice. *Journal of Consumer Research* 39(5), pp. 899–917.

Beljaars, D. and Bervoets, J. (2021). 'The production of Tourette syndrome: Erasures, framings, and Silences', *Northern Network of Medical Humanities Congress 2021*, April 22–24, 2021, Durham/Virtual.

Belk, R., Seo, J. and Li, E. 2007. Dirty little secret: Home chaos and professional organisers. *Consumption, Markets and Culture* 10(2), pp. 133–140.

Bell, S.L., Foley, R., Houghton, F., Maddrell, A. and Williams, A.M. 2017. From therapeutic landscapes to healthy spaces, places and practices: A scoping review. *Social Science & Medicine* 196, pp. 123–130.

Bergson, H. 1911 [1896], *Matter and Memory*, trans. Paul, N.M. and Palmer, W.S., London: George Allen and Unwin.

Bervoets, J. and Beljaars, D. in review. From deficit to surplus models of mental illness. Tourette Syndrome: A case study. *Social Science & Medicine*.

Bissell, D. 2011. Thinking habits for uncertain objects. *Environment and Planning A* 43, pp. 2649–2665.

Bodeck, S., Lappe, C. and Evers, S. 2015. Tic-reducing effects of music in patients with Tourette's syndrome: self-reported and objective analysis. *Journal of the Neurological Sciences* 352(1–2), pp. 41–47.

Chouinard, V. 2012. Mapping bipolar worlds: Lived geographies of 'madness' in autobiographical accounts. *Health & Place* 18, pp. 144–151.

Conradson, D. 2005. Landscape, care and the relational self: Therapeutic encounters in rural England. Health and Place 11, pp. 337–348.

Davidson, J. 2003. *Phobic Geographies*. Aldershot: Ashgate.

De Caluwé, E., Vergauwe, J., Decuyper, M., Bogaerts, S., Rettew, D. and De Clercq, B., 2020. The relation between normative rituals/routines and obsessive-compulsive symptoms at a young age: A systematic review. *Developmental Review* 56, p.100913.

Dewsbury, J.D. 2011. The Deleuze-Guattarian assemblage: Plastic habits. *Area* 43(2), pp. 148–153.

Dewsbury, J.D. and Bissell, D. 2014. Habit geographies: The perilous zones in the life of the individual. *Cultural Geographies* 22(1), pp. 21–28.

Dion, D., Sabri, O. and Guillard, V. 2014. Home sweet messy home: Managing symbolic pollution. *Journal of Consumer Research* 41(3), pp. 565–590.

Duff, C. 2014. *Assemblages of health: Deleuze's empiricism and the ethology of life.* Dordrecht: Springer.

Gesler, W.M. 1992. Therapeutic landscapes: Medical issues in light of the new cultural geography. *Social Science & Medicine* 34(7), pp. 735–746.

Gorman, R. 2017. Therapeutic landscapes and non-human animals: The roles and contested positions of animals within care farming assemblages. *Social and Cultural Geography* 18(3), pp. 315–335.

Grosz, E. 2013. Habit today: Ravaisson, Bergson, Deleuze and us. *Body and Society* 19(2–3), pp. 217–239.

Haraway, D. 2016. *Staying with the Trouble: Making Kin in the Chthulucene*. Durham: Duke University Press.

Hayhoe, M. et al 2003. Visual memory and motor planning in a natural task. *Journal of Vision* 3(1), pp. 49–63.

Ihde, D. 2009. *Postphenomenology and Technoscience*. Albany, NY: SUNY Press.

Kirkham, P. 2017. The line between intervention and abuse – autism and applied behaviour analysis. *History of the Human Sciences* 30(2), pp. 107–126.

Land, M.F. et al. 1999. Eye movements and the roles of vision in activities of daily living: making a cup of tea. *Perception* 28, pp. 1311–1328.

Latimer, J.E. and Miele, M. 2013. Naturecultures? Science, affect and the non-human. *Theory Culture & Society* 30(7–8), pp. 5–31.

Latour, B. 2005. *Reassembling the Social: An Introduction to Actor-Network-Theory*. Oxford: Oxford University Press.

Law, J. 2004. *After Method: Mess in Social Science Research*. London: Routledge.

Lea, J. 2009. Post-phenomenological geographies. In: Kitchen, R. and Thrift, N. eds. *International Encyclopaedia of Human Geography*. London: Elsevier, pp. 373–378.

Lingis, A. 1995. The world as a whole. In Babich, B.E. ed. *From Phenomenology to Thought, Errancy, and Desire: Essays in Honor of William J. Richardson, S.J.*, Berlin: Springer Science+Business Media, pp. 601–616.

Lipsky, D. 2011. *From Anxiety to Meltdown: How Individuals on the Autism Spectrum Deal With Anxiety, Experience Meltdowns, Manifest Tantrums and How You can Intervene Effectively*, London: Jessica Kingsley Publishers.

Mcphie, J. 2019. *Mental Health and Wellbeing in the Anthropocene. A Posthuman Inquiry*. Singapore: Palgrave Macmillan.

Mol, A. 2002. *The Body Multiple: Ontology in Medical Practice*. Durham, NC: Duke University Press.

Owen, J. 2018. *'Out of sight, out of mind' – The place of self storage in securing pasts, ordering the present and enabling futures*. PhD Thesis, Cardiff University.

Rosenblatt, A. 2018. Autism, advocacy organizations, and past injustice. *Disability Studies Quarterly* 38(4), pp. np.

Schatzki, T.R. 2008. *Social Practices. A Wittgensteinian Approach to Human Activity and the Social*. Cambridge: Cambridge University Press.

Tyner, J. 2016. *Violence in Capitalism: Devaluing Life in an Age of Responsibility*. Lincoln, NE: UNP – Nebraska.

Wilson, K. 2003. Therapeutic landscapes and First Nations peoples: an exploration of culture, health and place. *Health and Place* 9, pp. 83–93.

Wong, S. and Beljaars, D. forthcoming. Geographies of disability: on the potential of mixed methods, Rosenberg, M.W., Lovell, S., and Coen, S.E. eds. *Routledge Handbook of Methodologies in Human Geography*. London & New York: Routledge. np.

9 A compulsive worlding of (post)humanity

The development of compulsivity as an environmental, experiential phenomenon with a distinct socio-spatial existence raises several questions. This chapter takes a step back and considers what kind of questions are conjured up, and it explores the implications for a variety of subsequent knowledge systems and knowledge creation practices. This is not an exhaustive list of questions, nor does it provide all the answers. Rather, this chapter extends the logic built up in the previous chapters through the introduction of new strands of inquiry to point out how and where the geographies of compulsion and compulsive geographies should be accounted for. Questions regarded as of particular significance are those pertaining to the relevance of expansion of the argument beyond people with a Tourette syndrome diagnosis, subjectivity and subjectification processes, liberal humanism, and legal-ethical questions. Elsewhere I raise more immediate questions related to the geographical study of disability and health (Beljaars 2020, Wong and Beljaars forthcoming). Exploring these questions is not a way to rationalise a certain 'irrational' type of human action. Rather, it sets out to create a new space for empathy for such action and reinvigorate a new way of caring for those who are struggling the most with compulsivity but also for anyone of us in moments of compulsion.

Compulsivity names an openness to the world: it is one of the traceable ways in which the extracorporeal, the *new*, grabs hold of bodies. Echoing Deleuze's vitalist rendering of repetition (1994 [1968]), even seemingly recurring compulsive interactions are always new; not only because every time they will be slightly different, but also as despite being expected they cannot be 'pre'-performed[1]. Underpinning this openness to the world is compulsivity manifesting as a creative faculty of life. Hence why behaviourism as a mode of explanation cannot wholly capture compulsivity as it seeks striation through recurring patterns in multiplicity of situations. Rather than recognising how situations crucially underpin prioritisation in experiential life, these strongly reductive accounts are then attributed to a central ego that fails to inhibit responding to impulses or chooses the wrong one to respond to. Behaviour as a construct that is often used to pinpoint compulsivity leaves out important nuance and tends to gloss over surprise and

DOI: 10.4324/9781003109921-10

situational particularity; the 'new' cannot be granted entry into the understanding of compulsions and the conceptualisation of compulsivity.

The new is welcomed in the book's analysis of compulsivity through the mutual configurational process between the corporeal and the extracorporeal. Bodies navigate becoming part of body-object-space compositions, through which they configure into a compulsive interaction, and unbecoming part of them, which often means becoming part of one or multiple other compositions depending on the environment. Therefore, compulsivity challenges accounts of human-world relations that striate the world in accordance with the human. Instead, compulsivity as spatial phenomenon demonstrates how humans are intimately related to the world but do not dominate the interactions with it. Rather, for people with strong compulsive tendencies, this might be quite the contrary; they seem to live in a world of which its composites incessantly tear them away from intentional life.

Compulsivity is then reimagined as a multiplicity of ecologically situated compositions of body-object-space configurations that move with the body and become more and less intense with the perception of situations. The body with compulsive sensibilities can then be envisioned to be drawn into compulsive interactions to instate, restore, or not disturb these compositions. Therefore, compulsivity is a formative and organisational force that does not quite belong to the human or the non-human, but rather 'locates' in between. It can be conceived to emit from volatile body-object-world configurations and 'move' humans through the flesh (Wylie 2006, Hoel and Carusi 2015), and constitute a pre-modal perceptual understanding of time and space (Lingis 2000, Massumi 2002, Anderson and Harrison 2010). Compulsivity should therefore be seen as a 'character' of the body-world relationship beside a tribulation of the subject. To speak of a *compulsive humanity* is then a *contradiction in terms*. It is more accurate to speak of a *compulsive worlding of humanity*. To do so also means that the more intentional, moral, and idealistic aspects of human life are always underpinned by such worlding, despite playing out in ostensibly different domains of life. As such, this final chapter makes the case for a rehumanisation of compulsivity and a rehumanisation through compulsivity.

Compulsivity beyond pathology

Given that locating compulsivity away from the individual person and, as subjected to configurational withdrawal, even the body, makes its attachment to *certain* bodies challengeable. As such, it is possible to consider how thinking compulsive spatiality is relevant for people who do not have, or would not be eligible for, a Tourette's diagnosis. Such an endeavour begs the question of what value there is in recasting the circumstances of unpathologised human life as variably, but inherently, compulsive. It allows for better attunement to and stretches the kind of empathy and understanding that tends to remain reserved for non-pathological 'healthy' or 'normal' lives towards pathologised movement. Reflecting Tyler (2020), who argues that troubling categories allow for reclaiming the humanity that categorisation takes away, it breaks the victimisation, infantilisation, and stigma that such a movement tends to attract for people with a Tourette syndrome diagnosis.

Inflecting the extension of compulsive capacities to a broader humanity onto people with Tourette syndrome also brings to light why these people are so well-placed to discuss their experiences with compulsivity. Individuals who perform compulsions and other tics, in particular those with a Tourette syndrome diagnosis, are precisely those people whose minute bodily actions are scrutinised relentlessly; both by themselves and by familiar others. Often this started already at a young age that are formative years for self-reflection. The period before their diagnosis and eventual pathologisalisation required them not only to be hyperaware of their movements as these could be – and have been – questioned at all times. They also had to have an explanation ready for *every* move they make, which often includes both the general purpose of movement and, after being diagnosed, whether "it was Tourette's or 'themselves'". This group is therefore very well versed in knowing and putting into words what they experience in terms of compulsivity and how they are compulsive in their everyday lives. People whose actions are not scrutinised to that fairly extreme extent will probably be more forgetful about what they are doing exactly. Hence, any clinical, biological, social, or spatial difference between diagnosed and undiagnosed people is already mediated in life experience and language.

There are several entangled reasons for and approaches to blurring pathological boundaries on the matter of compulsive interactions. In terms of clinical populations, a commonly used 'fact' about Tourette syndrome is that only 50% of all people with Tourette syndrome have a diagnosis. Accordingly, the other 50% are part of the unpathologised population but belong to the clinical population about which neuropsychiatric claims are made. The discrepancy is blamed on eligible people not seeking help and on misdiagnosis through ignorance on the part of primary healthcare providers. As Dollnick (2007) argues, compulsions are easily interpreted as acts that signify persistence and skills that can lead to successes in life, which happened to professional basketball player Mahmoud Abdul-Rauf. Furthermore, the 'subclinical' population outside the Tourette syndrome diagnosis and the other 50% also "exhibit the whole range of repetitive phenomenology performed by GTS and OCD patients, and only differ quantitatively from them" (Cath et al. 2001: 181). This is a common stance, as it is acknowledged that children in particular perform tics and compulsions (see for instance Rachman and de Silva 1978, Salkovskis and Harrison 1984, Kurlan et al. 1994, Muris et al. 1997).

Further blurring the lines between the clinical and subclinical population, in line with the psychological model of mental illness (Freeman and Freeman 2013), Szatmari (2004) argues that bodily interactions that are declared as symptomatic occur along a continuum (also see Williams 2005). For instance, having particularly strong preferences for the organisation of one's work or study space, kneading the tube of toothpaste in a particular way, or 'running' along the car and jumping over trees and lampposts in one's imagination when looking out of a moving car. Psychoanalytical models of compulsions and tics that have now been made redundant in the psychological analysis of Tourette syndrome, but are nevertheless relevant to this discussion, also suggest that the diagnosis signifies an extreme version of what is declared 'normal behaviour'. Ferenczi (1921) for instance alludes to compulsive behaviour through terms such as "exaggeration" and "enhancement" of 'normal' behaviours. He also argues that stereotypies are 'simpler' tics, which are also associated with diagnoses of autism spectrum, phobia, and anxiety disorders.

Seeing compulsivity as a process that involves also unpathologised bodies on their sensory and performative faculties and finding a biological, material 'anchor' in these bodies requires a reframing of, and a shift in, the pursuit of the materialist, biological research. If compulsive interactions have to be performed whether people recall doing them or not, and if they are unrelated to the self, and serve no purpose beyond finishing the interaction, could they not be the result of biological processes such as hunger, thirst, and sexual drive? Halting the assumption of the biological processes of compulsivity as inherently problematic for a moment allows for considering compulsivity along the same lines as the accepted biological processes that sustain bodies and people in their survival. Drew Leder (1990: 21) sees hunger, thirst, and sexual drive as corporeal states and "modes whereby the environment stands forth", and that "color the perceived world". Increasing hunger, and in mediation of food sensitivities, the world "channel[s] attention and activity toward potential sources of gratification" in the form of food (ibid.). In compulsive terms, the bodily surroundings would become increasingly perceived as 'intriguing' in Sage's words and having the potential to be interacted with until gratified.

Biological processes of hunger and thirst are rendered normal and studied for the biological effects of different food stuffs and fluids are understood to be beneficial for the person. Comparing compulsivity to hunger and thirst implies that compulsivity would serve a goal beyond extinguishing itself, and therefore be an acceptable, non-pathological process that has been misinterpreted, ostracised, and eschewed in the neuropsychiatric, biomedical, and clinical sciences and broader societal realm. In this sense, compulsivity has not been studied along the same lines and outside its normative frame: whereas digestion studies do not only focus on the mouth, stomach, or gut but also on differences in food stuffs and excrement, studies of compulsivity tend to focus on the brain, to a lesser extent on the body parts where compulsions can be observed to happen, and rarely on the extracorporeal. Indeed, all bodily processes register in the brain, but that should not be the only focus. As mentioned in Chapter 2, research efforts are aimed at highlighting the ontological difference *of the phenomenon as such* from the normal, but do not consider ontological differences between what in linear processes is considered 'input' (i.e. that what is sensed and to which certain processes in the brain respond) and 'output' (i.e. what bodily movement and/or sound is performed based on the result of the brain processes). Thus, the narrow focus of the biological dimensions of compulsivity has cut up the process to such an extent that the compulsive process as a larger 'whole' may be analytically unrecognisable. Focussing on different dimensions of compulsivity – the whole body and its socio-material surroundings – could bring more insights.

As such, to varying extents, all people would have tendencies to compulsively interact with particular situations and may experience some sort of push to do so or discomfort when they decide not to do so. The emergent *compulsive geographies* would render visible the tiny in-between acts that people catch themself doing, for example, stepping on the middle of the flag-stones in the pavement. Such interactions often fail to leave a memory but can be useful and enjoyable (Goldman 2012). Fidgeting as one such category is found to be a successful kind of interaction to cope with stressful situations (Farley et al. 2013). This does not mean that one should use expressions such as 'being a bit OCD' when an undiagnosed person organises objects according to symmetricity or diligently cleans their kitchen countertop. It is also a misplaced gauging of the suffering that such acts can conjure up, in severity of distress, interruption of further life, and time lost to perform multiple, complex, potentially painful, and extensive interactions. Nonetheless, compulsivity can be suffered in silence and claims of 'being a bit OCD' can be made by people who do not have a diagnosis but fit the criteria. A situational, ecological existence of compulsivity does allow for recognising the compulsive pressure behind the performances that is expressed by undiagnosed people without deriding the severity and the hopelessness that is often experienced by diagnosed people. The absence of suffering does not equate to an absence of compulsivity.

Indeed, as pathologised people are more attuned to the perceptual directive forces of the surroundings than unpathologised populations, and more attuned to the immemorial call of the world that is beyond reason, they could tell us more about the world beyond reason; the ways the world captures us; how we live situated lives beyond rationality. As Manning and Massumi (2014: 11) remind us in relation to their scepticism about and problematisation of the 'epidemic' of 'attention-deficit disorder', "might not the diagnoses betray an inattention on the part of adults to an attentiveness of a different order? One mode of existence's deficit may be another's fullness". Rendering acts such as fidgeting that blur the boundaries between compulsive and other-than-compulsive interactions then emphasises "the inherent and continuous susceptibility of corporeal life to the unchosen and unforeseen" (Harrison 2008: 427). If we halt normative judgement of compulsive interactions then it becomes easier to see that it reveals a bodily connection with the surroundings that is delicate and real. Compulsivity as beyond pathology then organises spatialities, materialities, and sensibilities in such a way that it allows immediately phenomenological and/or observable types of body-world formations that can stay ephemeral in non-representational geographies of the body and embodiment of certain places.

On subjectivity, intentionality, and meaning

As a meshwork ecology of potentially violent body-object-space compositions that reigns bodies into interacting compulsively, compulsive embodiment provides new insights into the spatiality of subjectification processes. Whilst the compulsions arise within the formation of subjectivities that involve the body, they do not centre it; neither as an initiative for action, nor as a point from which perception emanates, as is invoked in geographies of the lifeworld (see Seamon 2017). The evolving landscape that is shaped by these multiple compositions makes it 'sticky' when moving through it, as entering new compositions still requires dealing with the effects of having been part of the compositions that the body left moments before. An imprint of the tip of the table that was just pressed into in the index finger cushion that is still felt, the minor elation and visual rendering of the 'perfect' curves of wood print in a door frame a body moved through, and the choreographed walk to 'step over' the lines that erupt from the corners of a room all linger. Affirming the asymmetry of subjectification processes (Deleuze and Guattari 2004 [1980], Braidotti 2006), compulsivity discloses how dynamic subject boundaries are formed by and stick to the boundaries of the compositions.

Compulsions like Ginny's interaction with the shopping cart, grocery bags, and the car indicate how compulsivity is not only beyond the psyche and beyond the human, but also how the human is not even always central to actions that involve the human body. Rather, the parameters of compulsive systems and the spatial reach of percepts work on the body and *subject the person* to comply with the necessity for (re)instating just-rightness. Any possible explanation of compulsive action can thus not only not be genuine, given it is not incited by the person who interacts, any insight can only be partial at best. If this proposition is accepted in conjunction with the impossibility of always being able to discern compulsive from other-than-compulsive action, following through with this in terms of research implications, how bodily movement is interpreted and analysed should be re-imagined as well. An inability to explain or describe bodily action should not *without question* be ascribed to an incapacity of the person involved; their bodily movement may have been instigated beyond them. Explanation of bodily action is simply not always possible, and attempts to force them become ingenuine.

The ways in which bodily action has been conceptualised and considered to inform insights into human life do not account for compulsivity – hence the pathologisation of its rendition as arising from the faulty brain of an individual person – in particular through intentionality. Action is always considered to be intentional because meaning can come from action. In broad terms, this is what practice theories emphasise (see Chapter 1). However, inspired by Alphonso Lingis' (1998, 2000) contentions that transform processes of perception *of* the extracorporeal to perception *with* it (see Rose 2006, Wylie 2006), this can open up analytical possibilities to place compulsive experiences as corporeal subordination. In turn, by destabilising the subject/object divide, the human subject is equally not intentional towards, and therefore not prior to, non-human objects (Zahavi 2003, Simonsen 2013, Ash and Simpson 2016); rather it is in their appearance that experience is constituted (Ash and Simpson 2016, McCormack 2017). As such, intentionality emerges in the direction the compulsion unfolds with. As the human is *involved with but does not instigate compulsions,* the accomplishment of compulsivity is not only relevant for the human, but also for the non-human. Compulsivity as a form of action that binds bodies with other bodies, objects, and spaces suggests that the intentionality that is subsumed under action is then a distributed and *double* more-than-human kind of intentionality.

Echoing postphenomenological and posthumanist arguments that seek to ontologise the non-human as an active part of life – human and other – despite its broad usage that suggests it to be a universal underpinning of human action, intentionality is a less useful concept to indicate how and why bodily action takes place. On the one hand, and indicating qualitative difference, using intentionality to indicate the direction of human action should be more critical. Employing it without question 'others' *non-* or *un*intentional movement, which skews social analyses of the body and quietly upholds *all* bodily movement wrongly as inherently and unwaveringly intentional. On the other hand, and indicating quantitative difference, using intentionality to analyse human action should be considered as always partial, as compulsivity names an ecologically constituted more-than-human that underpins all bodily movement to a varying extent.

Underpinned by a more-than-human type of intentionality, compulsivity then calls into question the relationship between performativity and signification. The idea in humanist approaches to personal geographies that bodily action also conjures up meaning has placed performativity firmly in the analytical focus. And whilst this focus has been very successful in highlighting how the fleeting, non-representational, and seemingly insignificant dimensions of everyday lives play a fundamental role in meaningful life as such, the position of meaning as be-all and end-all obscures dimensions of life such as compulsivity. Such a posthumanist analysis of a worlding reveals other dimensions where life plays out that shifts and slides underneath and falls in between meaningful ones. A social analysis that is not immediately guided by the pursuit of meaning, but that employs a framework of subjectivity, can then help to better understand how the bodily surroundings actively produce experiential life. Leaving a search for meaning as the first and foremost goal allows going beyond anthropocentric concerns.

Whilst the spatial rendition developed here moves away from meaning and signification as object of pursuit analytically, compulsivity undeniably does play a part in the production processes of meaning. Indeed, compulsivity is not a nihilist gateway to understanding everyday environments as meaningless. Rather, it amounts to a more comprehensive conception of the ways in which the surroundings gain entry into everyday meaningful life. Tracing compulsive renditions of life worlds provides a unique understanding of how the extracorporeal is encountered in the suspension between pure perception and the meaning of individuated objects and how the latter constructs further representational life. In compulsive life worlds, meaning is a distant consideration, almost an afterthought. Rather than a pervasive explanation of human bodily interactions with the non-human, in compulsive worlds meaning seems narrow, a constrictive, reductive grasp of constellations bursting with vibrancy, surrounding the body and demanding to be seen, touched, heard, smelt, and tasted. In compulsive moments, meaning enacts an artificial bridge that forcefully separates the human body and its constituencies. Indeed, compulsivity reveals how meaning clutches to material structures; its 'void' analytically attracts attribution of meaning as the formation of compulsivity allows viewing structures of meaning from the outside within the same material formations of the body and the extracorporeal. It pinpoints how objects themselves gain meaning that relates to certain aspects of life through representation and how they say something about the self. And it pinpoints how spaces gain meaning within a larger physical structure and in one's life and activities. What is missing from this structure and what compulsivity reveals is how the extracorporeal world gains meaning in its excessiveness and its rhythms.

Compulsivity is an ordering principle that articulates *with* instead of *from* the human, which emanates at the level of the pre-cognitive and pre-individual. Although the prefix of 'pre' suggests lineages towards the cognitive and individual, it is more accurate to say that the compulsive lives alongside the cognitive and individual and does not necessarily morph into either. Indeed, following Manning (2013) and Manning and Massumi (2014) on autistic sensibilities, compulsivity marks how objects are always moving back and forth between coherence as meaningful objects and coherence in formation with others. Compulsive body-world formation is a kind of undercurrent that supports, sustains, and sets the conditions upon which symbolic life takes place. These compulsive and other-than-compulsive 'currents' are in constant mutual affirmation. On the one hand, when the compulsive current is well adjusted to the symbolic one, all presences converge and people thrive (after Anderson 2012). In these situations, the world is experienced to not vibrate and objects remain 'still', reflecting Ginny's experience of produce on supermarket shelves. In situations where the currents are maladapted to each other the person 'falls still' and remains stuck (e.g. when Dylan cannot do anything he wants to do and lays on his bedroom floor).

On the other hand, suffering then emerges with experiences of autistic people when they do not quite know what a particular room is (Davidson 2010, Davidson and Henderson 2010), have trouble locating the self in delusional and manic episodes (Parr 1999, Chouinard 2012), where bodily boundaries dissolve in anxious states people with OCD sensibilities go through (Segrott and Doel 2004) and agoraphobia (Davidson 2000a, 2000b, 2003, Callard 2006), and with experiences of not feeling 'safe' in a place (Coyle 2004). Amidst the flow of these 'currents', compulsive interactions are then ways to align them; they are emergent from the situation where the currents increasingly become problematic, which is felt by the human as the urge. The compulsive interactions that follow are then restorative of the 'maladapted' currents. Nonetheless, as noted before, compulsions and their legacies can be enjoyable as the new post-compulsion situation allows one to thrive. It may be possible to think the just-right vitality that emerges from the alignment and reinforcement of these currents.

Some rehumanising politics

A vitalist understanding of compulsivity shifts its place in society and under governance to a politics of surplus, rather than of lack. It rebukes reductionist efforts, as these will necessarily be incomplete in their attendance to the political nature of compulsivity. Rather, it merits following Elizabeth Grosz's (2017) account that attends to the immateriality of the conditions of materiality and which she develops as a politico-ethical system. Whilst compulsivity hinges between bodies, objects, and spaces, precisely its formative performativity organises a politics; both as ephemerally emergent from compulsive situations, and in its implications as a particular incessantly anticipated type of bodily 'capture'. In this anticipation there is an ethical necessity to respond, and a need for political accommodation for actualisation of the performance.

In a world in which bodily action that is deemed 'normal' and 'healthy' because it is reasoned or emergent from a subject or self, compulsions betray the subject. This becomes particularly apparent in liberal humanisms that assume sovereignty of the human subject. In addition to many other affective dimensions of life, compulsions demonstrate that this is not always the case, and that fluctuations and kinds of sovereignty are difficult to determine, given that compulsivity is a slippery phenomenon in pinpointing when it intensifies. As such, they affect the political subject. Considering broader humanity as compulsive to varying degrees, then, has certain political and ethical consequences. The presence of capacities to be compulsive taints the rational political subject. As such, compulsivity flies in the face of multiple core ideas about what it means to be human in primarily Western cultures and value systems. Therefore, compulsivity offers a new critique of the rational, purposive, intentional, meaning-seeking human subject that underpins the political figure in liberal democracies and upon which much of the political inequality is built. It challenges views of the figure of the human as categorically separate from the non-human and overlapping views that see the corporeal as categorically different from the extracorporeal.

The way compulsivity challenges the human political subject extends to the ostensibly intertwined economic one; in particular the human subject under capitalism. Similar to the political subject, the economic subject under capitalism is rationalised, albeit not by the same set of rationalising principles. Rationality implies a predictability, and compulsive action is an inherently unpredictable phenomenon. Whilst in liberal democracies political subjects ought to pursue what is best for themselves – socio-economic, religious, and cultural markers should render the political subject predictable to a certain extent – capitalist logic also requires a predictability of the economic subject for the production of its desirous affects to latch onto. Hence why Deleuze and Guattari (2004 [1972]) posit the process of schizophrenia as challenging capitalism as well as state capture. Bissell (2011: 2656) elegantly captures the challenges that unpredictable, seemingly irrational, and excessive movement poses:

> These movements are also deemed to be problematic since they appear excessive. Governance regimes of intervention are often justified through anxiety over free-falling movements that have the potential to go beyond 'tipping points' of no return. Valorisation of 'sustainability' as a set of loose-knit imperatives to moderate, reign-in, and store often underpins such interventions underpinned by an 'ethic of restraint' (Doel 2009: 1054). In the eyes of capital these excessive bodies are unproductive when they put a strain on economic and political resources or become incapable of being susceptible to *other* movements, forces, and seductions.
>
> (Bissell, Thinking habits for uncertain objects, 2011, p. 2656, Emphasis original)

In compulsive moments that can be anticipated to erupt from *every body* in varying intensity, people cannot be predicted, nor trusted, and thus become dangerous. In order to prevent consumers to become more unpredictable, capitalist societies have no choice but to render compulsivity abject; a 'flaw' that requires 'rectification'.

Rectification becomes possible through pathologisation which permits various forms of intervention that would ostensibly normalise people. Biologising compulsivity as human defect of the brain could then be effective. The neuropsychiatric provision of material evidence of brain activity or neurobiological structure that is significantly different in pathologised people with compulsions allows extending the argument to the whole brain, which also implies a person's cognitive capabilities, and thus the whole person as they are a material unity. Whilst this motivates normalising practices that increase the predictability of these people, which, in turn, can be moni-tised, this allows for dismissing people as a whole as ill, unproductive, and even dangerous. The latter is reflected in the invocation of mental illness as an excuse for criminal acts, and police violence committed against people who are labelled as unstable when they perform what is deemed to be irra-tional behaviour. With the increasing political swing to the political right in Western countries and technological capture of bodies through various forms of surveillance that allow high-level scrutiny of bodily movement, compulsive movement stands out as irrational. What is understood as a rational, productive human figure seems to diminish rapidly; the minutia of such a figure and reflected in progressing brain research that finds rapid entry in public policy (see Gagen 2015, Pykett 2017, Whitehead et al. 2019).

As mentioned before, compulsivity exposes flaws in ideas of what it means to be human, which poses challenges to the humanisms with which profound decisions are made. Considering intentionality to underpin all human movement is a manifestation of the terms on which humanity is deemed exceptional in Western[2], techno-scientific, Judeo-Christian world-views. As the phenomenon of compulsivity immediately opposes intention-ality as emerging from the self, it undercuts the principles and values that presume such kind of intentionality as well. That includes universal notions of morality, which is not to say that people who are pathologised with com-pulsions have no sense of moral value or do not live by them. Rather, the distributed and ecologically constituted compulsive situations translate into momentary action that is not guided by *exclusively human* morality. In other words, compulsive interactions withdraw people from moral lives and inject them with amoral moments. By extension, a posthumanist con-ceptualisation of compulsivity questions if morality can ever be considered on human terms *only*. Expressed differently, moral choice cannot entirely belong to the human but needs to be extended to the situation. Compulsion is the precise demonstration of this requirement; it is not the opposite of acting morally, of acting without morals, but it questions the possibility of pure moral acts. Moral action is only ever possible to varying extents.

In line with posthumanist politico-ethical concerns (e.g. Haraway 2016, Weinstein and Colebrook 2017, Braidotti 2019), compulsivity could poten-tially be a way to help adapt humanism and retrieve it from its exceptionalist beliefs. Human exceptionalism, born out of the social hierarchy philosophies that (re)produce white supremacy, is based on technological, cognitive, and moral superiority over animals and other organisms has been used as justifi-cation for capitalist economic expansions that have led to the development of the agro-industrial industry, allows for ongoing earthly resource depletion, and produces widespread global pollution. Raising compulsivity as phenom-enon that suggests that humans are less sovereign because they are more dependent on their environment, and less capable of rational behaviour may help to diminish the articificially created degrees of difference between the human species and the nonhuman planet. Compulsive posthuman politics as a mode of being-with then revises where empathy should and can be cast. Indeed, it is a springboard for a kinder understanding of human bodies that hold as much power to resist the increasing political appropriation of bodies and their action by challenging rationalisation efforts as it does in creating new connections between bodies, as well as extending empathy between humans as well as between humans and animals and other organisms (after Haraway 2016). Struggles that such empathy unveils point to the necessity of fundamentally extending the political realm to the non-human (e.g. Sundberg 2014, Massey and Kirk 2015).

Ethical questions

A re-conception of compulsivity requires an expansion of the normative ontology from situated between different human bodies to between human bodies and their non-human environment. This has certain ethical implications that echo poststructuralist challenges to universalist, stabilised systems of relations overlapping with posthumanist concerns about the exclusion of the non-human. What follows is a brief exploration as to what these may entail and what a generous, productive, and kind ethical concern may look like: not one that posits compulsivity in opposition to morality and open it up to judgement. Rather, such an ethical system ought to have immanent qualities, an ecological constitution that guides how to live well in the humdrum of everyday life. Considerations can follow two directions; one that explores the ethics according to which compulsivity is shaped and compulsions are produced, and one that places the phenomenon in the social world and thinks through the consequences.

Firstly, such ethical principles should be tied to compulsivity's existence as moving between being inside and outside awareness, having varied and unpredictable presence, and requiring differing responses of the person. Indeed, a more-than-human, situational ethics is required that does not centre the subject as initiator nor as particularly vulnerable recipient to potentially violent outcomes. Rather, the ethical system that underpins compulsivity is guided by heterogeneously constituted *just-rightness* that acts as a certain incomprehensible truth, if only in its affirmative character and in its spatiotemporal rhythms. Returning to Alphonso Lingis, who argues that things have their own vibrancy and our response to their call makes us human (1998, 2000), he suggests that there is an ethics in the need to respond to what is perceived, even if these things do not have a face. Compulsion is a way of investigating this ethics: it brings to the fore how we attune to objects prior to their utilitarian capture. Just-rightness recasts a new order onto human and non-human presences in a situation that is actualised when the human body is reconfigured towards the compulsive situation and completes it.

To account for the ethical principles that underpin compulsions, the system of ethics that is based around situational change as developed by Alain

Badiou (2013) can be explored. As this system considers what happens in unfolding situations when the status quo is disrupted, the configurational occurrence in compulsivity resembles the rhythms of the re-making order. As mentioned earlier, compulsions interrupt the cultural realm, and have consequences for it, but do not take off from it. In this sense, the start of compulsive configuration constitutes the start of the disruption and achieving just-rightness in compulsions could be considered to constitute the 'truth event' that striates a new order. Simultaneously, if the person involved goes through with the configuration of their body and remains faithful to the truth event, the basis changes on which subjectification processes take place. This makes compulsivity an immediately more-than-human onto-ethical system, in which the more-than-human values govern the situation. Compulsive ethics thus contribute to efforts that aim towards truths that are incomprehensible to a person when they happen but have legacies that are primarily *for the non-human* (Braidotti 2006). Expanding Haraway's (2016) development of ethics of more specified care for the non-human, compulsion can then be regarded as a form of non-human caring; the interactions geared towards just-right optimisation might not be understandable from a human perspective, but without interpreting the legacies of individual compulsions, extracorporeal, non-human entities are addressed on their circumstances. Arguably, this makes compulsivity a form of response to more-than-human situations in which the human's needs are not prioritised.

The spatial underpinning of compulsivity that is developed in this book joins calls for a hesitation to assume that bodily action can be grasped in terms of behaviour. Rooted in deconstructionist approaches in the social sciences and humanities, and in critical feminist, postcolonial, and non-representational theories in human geography, the strong situatedness of compulsions further challenges the merit of universalisation of bodily action through behavioural reductionism and rigidification. Behaviour can only ever work if the origin and end circumstances of bodily action are the same, and it requires a singular and coherent self that rationalises the relations between action, purpose, and meaning. Behaviour as uncontextualised and a distinctly human construct purports that bodily action can be predicted and interpreted, and translated into morality and normativity. In turn, this allows for making claims about the person and holds a truth about the moral goodness of the person. In times that are characterised by a crisis in truth and fake news[3], and words become increasingly compromised, bodily action supposedly reflects another, ostensibly more believable, truth. However, compulsivity demonstrates that this amounts to a fallacious interpretation, except if a more-than-human truth is accepted or one that is fleeting. At the same time, behaviourism opens the door for normative judgement of the person who performs compulsions, which, in turn, exposes them to different forms of violence.

Secondly, focussing on the more-than-human circumstances of the individual, ethics that underpin compulsivity challenge neoliberal choice perspectives as well as behaviour as a concept of knowing, policing, and governing human behaviour on the individual and momentary level. Additionally, in the affirmative mode such ethical considerations underpin the possibilities to relate to, care for, and take in the pain of others (Levinas 1999, Puig de la Bellacasa 2017, Braidotti 2019 [1974]). Whilst compulsions are not reactions to social processes, they are constituted with(in) the social world and have social consequences, in particular if they involve other human, animal, or plant bodies, are (self)destructive, or seemingly challenge social norms (Davis et al. 2004, Buckser 2006, 2008). In this sense, ethical principles that support and shape compulsions have consequences for other ethical systems: they are tied in with the Western, liberal, universal human-rights type of ethics.

A distributed ethics that such a shift entails (see Coyle 2006) raises political questions about ownership and governmentality of the conditions and moral responsibility for the act. Reconceiving compulsivity as a more-than-human condition requires the reframing of ongoing debates on the location of stigma and shame (see Woods 2017). Instead of situating all ownership with the person performing the act (Bervoets forthcoming), it would take away any ground for stigma as stigma targets people, not acts. Indeed, any blame that is now cast onto the human would then become diffused over the situation. This acknowledges the powerlessness of the experience of having to perform compulsive acts that are humiliating. Such an acknowledgement is not only long overdue, it is also necessary to put in place sooner rather than later.

Imogen Tyler's (2020) critical appraisal of stigma as a practice of power and form of violence creates a foreboding future. In the absence of in-person meetings, and in the recorded curation of bodily movement through social media, bodily action has arguably become under even more scrutiny. Indeed, during the first year of the Covid-19 pandemic, a burgeoning number of people – mostly teenage girls – are reported to be unable to stop ticcing (Heyman et al. 2021). Being more visible than ever before, compulsions and tics seem to become a phenomenon that may increasingly be rendered abject, thereby punishing the person who performs them, which opens people up to social and state violence.

Thinking through the consequences of accounting for a compulsive worlding of the wider population primarily in the ethico-political context of Western societies, some socio-political differences can be identified. Based on the 'libidinal economy' that sustains compulsivity, living and residing in cluttered spaces is particularly demanding as they are more likely to be imbued with many compulsive systems that need stabilisation. Therefore, living in small spaces with many possessions that cannot be readily put out of sight is more challenging than living in larger houses that are more flexible in their organisation of things. Furthermore, living in places that can be altered to a greater extent; a privilege that comes with owning one's home and having the means to make changes, is less challenging than living in a place that is restricted in its adaptability, which tends to be a feature of rental homes, as well as shared homes, including living with one or more parental figures. Furthermore, working or being educated in noisy, rapidly changing situations that cannot be altered is also likely to be very tiring because of the compulsive dimension of life; the less compulsive systems can be stabilised, the more toll it may take on the energy and emotional tolerance. Many service jobs and manual labour jobs fit this category, as do mass in-person classes, workshops, and physical education. These two factors, housing and work/education circumstances, overlap with people with lower incomes and various kinds of students. With economic struggles, compulsive struggles could therefore be conceived to affect life in more profound ways.

Notes

1 Pre-performed in the sense of performed when chosen to do so at a convenient time to avoid disruption to other-than-compulsive life.
2 This is not to say that other cultures are necessarily more welcoming, but that Western societies are ideologically and systematically more hostile.
3 Whether this development is deemed a crisis will depend on the position of universality and authenticity in cultural value systems, which will vary across societies. As such, the naming of it as crisis is primarily an interpretation that is held in Western societies.

References

Anderson, B. and Harrison, P.eds.2010. *Taking-Place: Non-Representational Theories and Geography*. Farnham: Ashgate.

Anderson, J. 2012. Relational places: The surfed wave as assemblage and convergence. *Environment and Planning D: Society and Space* 304, pp. 570–587.

Ash, J. and Simpson, P. 2016. Geography and post-phenomenology. *Progress in Human Geography* 40(1), pp. 48–66.

Badiou, A. 2013. *Ethics. An Essay on the Understanding of Evil*. London & New York, NY: Verso.

Beljaars, D. 2020. Towards compulsive geographies, *Transactions of the Institute of British Geographers* 45, pp. 284–298.

Bervoets, J. forthcoming. Tourette Syndrome and Dynamic Moral Responsibility: a healthier look at Tourette's?, PhD Thesis, University of Antwerp.

Bissell, D. 2011. Thinking habits for uncertain objects. *Environment and Planning A* 43, pp. 2649–2665.

Braidotti, R. 2006. *Transpositions: On Nomadic Ethics*. Cambridge: Polity Press.

Braidotti, R. 2019. *Posthuman Knowledge*. Cambridge: Polity Press.

Buckser, A. 2006. The empty gesture: Tourette syndrome and the semantic dimension of illness. *Ethnology* 45, pp. 255–274.

Buckser, A. 2008. Before your very eyes: Illness, agency, and the management of Tourette syndrome. *Medical Anthropology Quarterly* 222, pp. 167–192.

Callard, F. 2006. The sensation of infinite vastness; or, the emergence of agoraphobia in the late 19th century. *Environment and Planning D: Society and Space* 24, pp. 873–889.

Cath, D.C. et el. 2001. Repetitive behaviors in Tourette's syndrome and OCD with and without tics: what are the differences? *Psychiatry Research* 101, pp. 171–185.

Chouinard, V. 2012. Mapping bipolar worlds: Lived geographies of 'madness' in auto-biographical accounts. *Health & Place* 18, pp. 144–151.

Coyle, F. 2004. Safe space' as counter-space: Women, environmental illness and 'corporeal chaos'. *The Canadian Geographer/Le Geographe Canadien* 48(1), pp. 62–75.

Coyle, F. 2006. Posthuman geographies? Biotechnology, nature and the demise of the autonomous human subject. *Social and Cultural Geography* 7, pp. 505–523.

Davidson, J. 2000a '...the world was getting smaller': Women, agoraphobia and bodily boundaries *Area* 32, pp. 31–40.

Davidson, J., 2000b. A phenomenology of fear: Merleau-Ponty and agoraphobic life worlds. *Sociology of Health and Illness* 22, pp. 640–660.

Davidson, J. 2003. *Phobic Geographies*. Aldershot: Ashgate.

Davidson, J, 2010. 'It cuts both ways': A relational approach to access and accommodation for autism. *Social Science and Medicine* 70, pp. 305–312.

Davidson, J. and Henderson, V.L. 2010. Travel in parallel with us for a while': Sensory geographies of autism. *The Canadian Geographer/Le Geographe Canadien* 54(4), pp. 462–475.

Davis, K.K., Davis, J.S. and Dowler, L. 2004. In motion, out of place: The public spaces of Tourette syndrome. *Social Science and Medicine* 59, pp. 103–112.

Deleuze, G. 1994 [1968]. *Difference and Repetition*, trans. Patton, P.R., New York, NY: Columbia University Press.

Deleuze, G. and Guattari, F. 2004 [1972]. *Anti-Oedipus: Capitalism and Schizophrenia*, trans. Massumi, B., London: Bloomsbury.

Deleuze, G. and Guattari, F. 2004 [1980]. *A Thousand Plateaus. Capitalism and Schizophrenia*. London: Continuum.

Doel, M.A. 2009. Miserly thinking/excessful geography: From restricted economy to global financial crisis. *Environment and Planning D: Society and Space* 27, pp. 1054–1073.

Dollnick, E. 2007. *Madness on the Couch: Blaming the Victim in the Heyday of Psychoanalysis*. New York, NY: Simon & Schuster.

Farley, J., Risko, E.F. and Kingstone, A. 2013. Everyday attention and lecture retention: The effects of time, fidgeting, and mind wandering. *Frontiers in Psychology* 4, pp. 1–9.

Ferenczi, S. (1921). Psycho-analytical observations on tic. *International Journal of Psychoanalysis* 2, pp. 1–30.

Freeman, D. and Freeman, J. 2013. *The classification of mental illness*. Retrieved July, 8th, 2017, from https://blog.oup.com/2013/05/classification-mental-illness-dsm-5-psychiatry-psychology-sociology/

Gagen, E. 2015. Governing emotions: Citizenship, neuroscience and the education of youth, *Transactions of the Institute of British Geographers* 40, pp. 140–152.

Goldman, D. 2012. *Our Genes, Our Choices: How Genotype and Gene Interactions Affect Behavior*. Cambridge, MA: Academic Press.

Grosz, E. 2017. *The Incorporeal. Ontology, Ethics, and the Limits of Materialism*. New York, NY: Columbia University Press.

Haraway, D. 2016. *Staying With the Trouble: Making Kin in the Chthulucene*. Durham: Duke University Press.

Harrison, P. 2008. Corporeal remains: Vulnerability, proximity and living on after the end of the world. *Environment and Planning A* 40, pp. 423–445.

Heyman, I., Liang, H. and Hedderly, T. 2021. COVID-19 related increase in childhood tics and tic-like attacks. *Archives of Disease in Childhood* 106, pp. 420–421.

Hoel, A.D. and Carusi, A. 2015. Thinking technology with Merleau-Ponty. In: Verbeek, P.-P. and Rosenberger, R. eds. *Postphenomenologial Investigations: Essays on Human-Technology Relations*. Lanham: Lexington Books, pp. 39–56.

Kurlan, R., Whitmore, D., Irvine, C., McDermott, M.P. and Como, P.G. 1994. Tourette's syndrome in a special education population: a pilot study involving a single school district. *Neurology* 44(4), pp. 699–702.

Leder, D. 1990. *The Absent Body*. Chicago, IL: University of Chicago Press.

Levinas, E. 1999 [1974]. *Otherwise Than Being, or, Beyond Essence*, trans. Alphonso, L., Pittsburgh, PA: Duquesne University Press.

Lingis, A. 1998. *The Imperative*. Bloomington, IN: Indiana University Press.

Lingis, A. 2000. *Dangerous Emotions*. Berkeley, CA: University of California Press.

Manning, E. 2013. *Always More Than One: Individuation's Dance*. Durham: Duke University Press.

Manning, E. and Massumi, B. 2014. *Thought in the Act: Passages in the Ecology of Experience*. Minneapolis, MN: University of Minnesota Press.

Massey, A. and Kirk, R. 2015. Bridging indigenous and Western sciences: research methodologies for traditional, complementary, and alternative medicine Systems. *SAGE Open* 5(3), np.

Massumi, B. 2002. *Parables for the Virtual: Movement, Affect, Sensation*. Durham and London: Duke University Press.

McCormack, D.P. 2017. The circumstances of post-phenomenological life worlds. *Transactions of the Institute of British Geographers* 42(1), pp. 2–13.

Muris, P., Merckelbach, H. and Clavan, M. 1997. Abnormal and normal compulsions. *Behaviour Research and Therapy* 35(3), pp. 249–252.

Parr, H. 1999. Delusional geographies: The experiential worlds during madness/illness. *Environment and Planning D: Society and Space* 17, pp. 673–690.

Puig de la Bellacasa, M. 2017. *Matters of Care. Speculative Ethics in More than Human Worlds*. Minneapolis, MN: Minnesota University Press.

Pykett, J. 2017. Geography and neuroscience: Critical engagements with geography's 'neural turn'. *Transactions of the Institute of British Geographers* 43(2), pp. 154–169.

Rachman, S. and de Silva, P. 1978. Abnormal and normal obsessions. *Behaviour Research and Therapy* 16(4), pp. 233–248.

Rose, M. 2006. Gathering 'dreams of presence': A project for the cultural landscape. *Environment and Planning D* 24, pp. 537–554.

Salkovskis, P., and Harrison, J. 1984. Abnormal and normal obsessions – a replication. *Behavior Research and Therapy* 22, pp. 549–552.

Seamon, D. 2017. *Life Takes Place: Phenomenology, Lifeworlds, and Place Making*. London: Routledge.

Segrott, J. and Doel, M.A. 2004. Disturbing geography: Obsessive-compulsive disorder as spatial practice. *Social and Cultural Geography* 5(4), pp. 597–614.

Simonsen, K. 2013. In quest of a new humanism: Embodiment, experience and phenomenology as critical geography. *Progress in Human Geography* 37(1), pp. 10–26.

Sundberg, J. 2014. Decolonizing posthumanist geographies. *Cultural Geographies* 21(1), pp. 33–47.

Szatmari, P. 2004. *A Mind Apart: Understanding Children With Autism and Asperger Syndrome*. New York and London: The Guilford Press.

Tyler, I. 2020. *Stigma: The Machinery of Inequality*. London: Zed Books.

Weinstein, J. and Colebrook, C. eds.2017. *Posthumous Life: Theorizing Beyond the Posthuman.* New York, NY: Columbia University Press.

Whitehead, M., Jones, R., Lilley, R., Howell, R. and Pykett, J. 2019. Neuroliberalism: Cognition, context, and the geographical bounding of rationality. *Progress in Human Geography* 43(4), pp. 632–649.

Williams, D.W. 2005. *Autism: An Inside-Out Approach: An Innovative look at the Mechanics of Autism and Its Developmental Cousins.* London and Philadelphia: Jessica Kingsley Publishers.

Wong, S. and Beljaars, D. forthcoming. Geographies of disability: On the potential of mixed methods. In: Rosenberg, M.W., Lovell, S., and Coen, S.E. eds. *Routledge Handbook of Methodologies in Human Geography,* London & New York: Routledge. np.

Woods, A. 2017. On shame and voice-hearing. *Medical Humanities* 43(4), pp. 251–256.

Wylie, J. 2006. Depths and folds: On landscape and the gazing subject. *Environment and Planning D: Society and Space* 24, pp. 519–535.

Zahavi, D. 2003. *Husserl's Phenomenology.* Stanford, CA: Stanford University Press.

Index

Printed in the United States
by Baker & Taylor Publisher Services